MW00860831

BRIGHT GALAXIES
DARK MATTERS

Masters of Modern Physics

Advisory Board

Dale Corson, Cornell University
Samuel Devons, Columbia University
Sidney Drell, Stanford Linear Accelerator Center
Herman Feshbach, Massachusetts Institute of Technology
Marvin Goldberger, University of California, Los Angeles
Wolfgang Panofsky, Stanford Linear Accelerator Center
William Press, Harvard University

Published Volumes

The Road from Los Alamos by Hans A. Bethe
The Charm of Physics by Sheldon L. Glashow
Citizen Scientist by Frank von Hippel
Visit to a Small Universe by Virginia Trimble
Nuclear Reactions: Science and Trans-Science by Alvin M. Weinberg
In the Shadow of the Bomb: Physics and Arms Control
    by Sydney D. Drell
The Eye of Heaven: Ptolemy, Copernicus, and Kepler
    by Owen Gingerich
Particles and Policy by Wolfgang K. H. Panofsky
At Home in the Universe by John A. Wheeler
Cosmic Enigmas by Joseph Silk
Nothing Is Too Wonderful To Be True by Philip Morrison
Arms and the Physicist by Herbert F. York
Confessions of a Technophile by Lewis M. Branscomb
Making Waves by Charles H. Townes
Of One Mind: The Collectivization of Science by John Ziman
Einstein, History, and Other Passions by Gerald Holton
Bright Galaxies, Dark Matters by Vera Rubin
Atomic Histories by Rudolf E. Peierls
Essays of a Soviet Scientist by Vitaliĭ I. Gol'danskiĭ
This Gifted Age: Science and Technology at the Millenium
    by John H. Gibbons

BRIGHT GALAXIES
DARK MATTERS

VERA RUBIN

Department of Terrestrial Magnetism
Carnegie Institution of Washington
Washington, DC

Springer

Library of Congress Cataloging-in-Publication Data
Rubin, Vera C., 1928–
    Bright galaxies, dark matters / Vera Rubin.
        p.   cm. — (Masters of modern physics)
    Includes bibliographical references and index.
    ISBN 1-56396-231-4
    1. Galaxies.   2. Dark matter (Astronomy).   3. Astronomers.
    I. Title.   II. Series.
QB857.R83   1996   96-25122
523.1´12—dc20   CIP

This book is volume seventeen of the Masters of Modern Physics series.

Printed on acid-free paper.

© 1996 Springer-Verlag New York, Inc.
All rights reserved. This work may not be translated or copied in whole or in part without the written permission of the publisher (Springer-Verlag New York, Inc., 175 Fifth Avenue, New York, NY 10010, USA), except for brief excerpts in connection with reviews or scholarly analysis. Use in connection with any form of information storage and retrieval, electronic adaptation, computer software, or by similar or dissimilar methodology now known or hereafter developed is forbidden.
The use of general descriptive names, trade names, trademarks, etc., in this publication, even if the former are not especially identified, is not to be taken as a sign that such names, as understood by the Trade Marks and Merchandise Marks Act, may accordingly be used freely by anyone.

Printed and bound by United Book Press, Inc., Baltimore, MD.
Printed in the United States of America.

9 8 7 6 5 4 3 2

ISBN 1-56396-231-4 Springer-Verlag New York Berlin Heidelberg   SPIN 10647189

for Bob
for Ruth
who are always there

CONTENTS

PART III
MATTER AND MOTION **95**

PART IV
THE ASTRONOMICAL LIFE: WOMEN IN SCIENCE
AND OTHER HEROES, COLLEAGUES, AND FRIENDS **151**

About the Series

Masters of Modern Physics introduces the work and thought of some of the most celebrated physicists of our day. These collected essays offer a panoramic tour of the way science works, how it affects our lives, and what it means to those who practice it. Authors report from the horizons of modern research, provide engaging sketches of friends and colleagues, and reflect on the social, economic, and political consequences of the scientific and technical enterprise.

Authors have been selected for their contributions to science and for their keen ability to communicate to the general reader—often with wit, frequently in fine literary style. All have been honored by their peers and most have been prominent in shaping debates in science, technology, and public policy. Some have achieved distinction in social and cultural spheres outside the laboratory.

Many essays are drawn from popular and scientific magazines, newspapers, and journals. Still others—written for the series or drawn from notes for other occasions—appear for the first time. Authors have provided introductions and, where appropriate, annotations. Once selected for inclusion, the essays are carefully edited and updated so that each volume emerges as a finely shaped work.

Masters of Modern Physics is overseen by an advisory panel of distinguished physicists. Sponsored by the American Institute of Physics, a consortium of major physics societies, the series serves as an authoritative survey of the people and ideas that have shaped twentieth-century science and society.

PREFACE

As far back as I can remember, I was puzzled by the curious workings of the world, and especially of the sky. My earliest recollection is of sitting in the back of a car at night, driving home along the Wissahicken Drive in Philadelphia, and asking my father why the moon is going where we are going. As we traveled, the bushes, trees, and distant hills all passed by, but the moon sat steadily outside my window. How did the moon know that we were returning from Bubba Cooper's house in West Philadelphia to our home in Mt. Airy, Pennsylvania? Undoubtedly, I was given an answer; probably I did not understand it. But I can still recall the excitement of the question. And now, some 60 years later, the excitement I get from asking questions of Nature is no less.

As a youngster, other questions followed, many puzzling for years. Why did the pictures on my bedroom wall jump back and forth on either side of my finger held in front of my face as I lay in bed blinking my eyes? How did drops of water in a flowing stream know on which side of a rock to pass? Could I, a lazy child, devise a street on which one sidewalk went uphill and one side went downhill, so that I could always walk downhill? These were my puzzles at age 3, or 5, or so. Sometimes I voiced them aloud, but not often. Thinking back now, I believe that even then I was more interested in the question than in the answer. I decided at an early age that we inhabit a very curious world.

As I grew older and continued to watch the sky, the questions became more conventional, even dull. Some I could puzzle out for myself. How many license plates can be made with three numbers and two letters? This puzzle I solved as we drove from Philadelphia to our new home in Washington, D.C. In those depression years, astronomy questions were answered by library books; other far-out questions were raised, not answered, in inspiring books by Sir James Jeans. Family friends Goldie and Mike drove Ruth (my sister) and me in their open car to the dark Virginia countryside and named the stars and the aligned planets; my parents (Rose and Pete Cooper) took us to Haines Point to see the spectacular aurora of 1939. These visual experiences were primary in causing me to become an astronomer. Some 40 years later, Goldie was to write to me "I've been thinking back to those days when we rode out under the stars. It couldn't be done now! Then, when you crossed the bridges into Virginia you were in the country. Now I feel it a nightmare to drive in Virginia even in the daytime."

> And so city children now grow up disconnected from the sky, and disconnected from yearly and monthly natural cycles. They do not know that every star they see belongs to our galaxy; that they live in a galaxy, whose disk forms the band on the sky we call the Milky Way. They fail to notice that the sun is high in the sky in some seasons and low in the sky in others. They do not recognize that the constellation of Orion introduces autumn, or that the planet Venus, the bright "star" in the twilight sky, will shortly disappear to arise as the morning star. When I visit Washington, D.C., elementary schools and ask who has seen the Milky Way, I get blank stares and shaking heads. Once, I actually got the response "she has," and several fingers pointed to the exceptional student. And the comment, "After I get injured, I might become a scientist," was one bright little boy's idea, in discussing his dreams for a career in sports.

> We are failing by not introducing school children to the wonders of Nature—the bugs, the rocks, the planets, the stars, and especially the unknowns. We are failing if children do not come to us with questions we cannot answer.

This book is a collection of 35 papers, essays, and talks, extending over 36 years. I have attempted to include the more general ones, often originally talks, rather than the more technical papers. Too many of them are recent. The combination of doing much work at home with a poor filing system, having a good memory, and using slides as a prompt, has ensured that notes for most of my early talks no longer exist.

Several themes recur in these works. The major theme is surely the study of motions of stars in galaxies and the evidence from these motions that most of the matter in the universe is dark. Such studies have been central to my career for more than 25 years. But other themes emerge as well. Our nearest (currently known) large spiral neighbor, the Andromeda galaxy (M31), is a galaxy I study over and over, as new telescopes and equipment offer new opportunities for learning more of its secrets. In this collection, the M31 galaxy appears in papers from 1973, 1987, and 1995. The careful reader may be able to trace our increased understanding of its complexity, which has come about from the combination of observations, theories, and computer simulations. It is not surprising that the study of the M31 galaxy led me to the studies of rotation of many other galaxies.

This book is mainly words—technical words, English words. But astronomy is a visual science, and discoveries can be made by looking. As Yogi Berra may have said, "You can see a lot just by looking." I have here included galaxy images. I hope the viewer will examine them carefully. Galaxies have their own personalities, just as people do. I hope some of their idiosyncrasies show. And in my talks I often refer to poems, songs, or paintings. A few of these are included too.

Scientists too seldom stress the enormity of our ignorance. Virtually everything we know about galaxies we have learned during the last 100 years. Galaxy spectroscopy will soon be 100 years old; a few of the major results are celebrated in "A Century of Galaxy Spectroscopy" in Part III of this book.

But what are the questions for future astronomers? What questions will astronomers be asking of the universe 100 years from now? A thousand years from now? Questions easy to list are the unsolved questions of the present. What is the age of the universe? What is the here-and-now rate of expansion of the universe? How much mass is there in the universe? What is the dark matter? Is its gravitational attraction sufficient to halt the expansion and to start the universe contracting? Do any nearby stars have planets harboring life, close enough so that we can communicate in a short time?

Then there are the questions we barely know enough to ask. A feeble list: Are there other universes? Will we ever communicate with them? How will our concept of the universe alter when gravitons are detected? As we peer into the universe we are peering into our past, but our "eyes" are weak and we have not yet seen to great distances. No one promised that we would live in the era that would unravel the mysteries of the cosmos. The edge of the universe is far beyond our grasp. Like Columbus, perhaps like the Vikings, we have peered into a new world and have seen that it is more mysterious and more complex than we had imagined. Still more mysteries of the universe remain hidden. Their discovery awaits the adventurous scientists of the future. I like it this way.

Winter Solstice, December 22, 1995. Altitude: 6100 ft; temperature: 10 °F. I write these words in a cozy log cabin along the Snake River in

Jackson Hole, Wyoming. This night the sun stops its southern migration and turns north, but it will be months before the North is warm. The animals we saw today—the moose bedded down in the sagebrush at the foot of the Tetons, the deer, the elk, and the trumpeter swans—all have millions of years of evolution that teaches them how to stay warm. As the sun sets high behind the snowy Tetons, geese are silhouetted against the yellow sky. Venus is brilliant, but Mars, Mercury, and the skinny crescent moon low in the west are hidden by the peaks. It's a small price to pay for the view of the mountains.

Most of the family (Dave, Judy, Karl, Allan, Michelle, Gene, Donna) and the grandchildren arrive tomorrow. Seven-year-old Eli especially will like to see the swans, for on his last visit to our house in Washington, D.C., he found a copy of *The Trumpet of the Swan* by E. B. White, with ALLAN RUBIN inscribed. Because it belonged to Allan as a child it is a special treasure. The other grandchildren—Ramona, Zan, and Laura—have seen the animals winter and summer, and they too know the treat of looking hard and spotting them.

After dark the sky is brilliant with stars. The Dipper stands on its handle at the horizon, with Cassiopeia opposite it past the North Star. In the city, we see pieces of the Dipper or Cassiopeia—never both at once. During this, our first night of this trip to Wyoming, it is easy to sleep; but also easy to get up periodically to see the Dipper rise and Cassiopeia fall. By morning of this longest night they have changed positions from the previous evening.

The cabin is a comfortable, warm place to write, but the outdoors has too many secrets we hope to view. Bob is already out. There is comfort, even inspiration, in these brief connections with Nature, that a professional city life permits only intermittently. It is a good compromise.

ACKNOWLEDGMENTS

I n the random walk that is life, my walk has probably been less random than many. In 1952, aged 23, I entered the Georgetown University graduate school to become an astronomer. I had a husband and a child and was pregnant with a second; I had a B.A. degree in astronomy from Vassar College, and an M.A. in astronomy from Cornell University. That I succeeded in two goals—to raise a family and to be an astronomer—is a tribute to many. Lee D. Gilbert (a Washington, D.C., Calvin Coolidge high school math teacher) forced me to think; Maud Makemson, Monica Healea, Lewis Feuer, Martha Stahr Carpenter, Hans Bethe, Richard Feynman, Phil Morrison, George Gamow, Francis J. Heyden, S. J., Carl Kiess, Martin McCarthy, S. J., and Charlotte Moore Sitterly all instructed and inspired.

Professional colleagues have been an important part of my life in science. My affectionate thanks to W. Kent Ford, Jr., John A. Graham, François Schweizer, Robert Herman, Donald Lynden-Bell, Don Osterbrock, Morton Roberts, Margaret and Geoffrey Burbidge, Norbert Thonnard, and the stream of post-docs who yearly remind me that science is an iterative process, that progress is made step-by-step, and that they are all luckier than I because they have recently been instructed in the latest advances, all of which I must learn myself. This is a lesson they too will shortly learn. Other colleagues, too numerous to mention, have offered enthusiastic support, and I thank them.

The Carnegie Institution of Washington has been of major importance throughout my professional career. Presidents Caryl Haskins, Phil Abelson, Jim Ebert, and Maxine Singer, and Department of Terrestrial Magnetism Directors Ellis Bolton, Tom Aldrich, George Wetherill, Louis Brown, and Sean Solomon actively supported my work, as have all my colleagues at DTM. I thank them. Above all, Merle Tuve was an inspiration during his Directorship. His piercing questions were almost daily, almost annoying, but always important and neglected at my peril.

I thank Walter Gruen for permission to reproduce the painting "Phenomenon of Weightlessness" by Remedios Varo, and Diane Ackerman for permission to quote from *A Natural History of the Senses*.

And a warm thanks to those who have made it work and have made it fun: Lucille Herring, Molly Schuchat, Janice Dunlap, Rosa Maria Esparza, Sandy Keiser, Michael Acierno, and especially Mary Coder, whose support never wavered. Above all, the family, wide ranging, extended. They know.

Part I
GALAXIES

Within a galaxy, everything moves. During two minutes spent reading these paragraphs, the earth has moved 2500 miles as it orbits the sun; the sun has moved 20,000 miles as it orbits the distant center of our galaxy. In a 70-year lifespan, the sun moves 300,000,000,000 miles. Yet, this vast path is only a tiny arc of a single orbit: it takes 200,000,000 years for the sun to orbit once about the galaxy. During the age of the solar system and of the galaxy, our solar system has made perhaps 50 orbits about the center. We live in a big, old galaxy—just one galaxy among billions in a big, old universe. In the universe, all galaxies are in motion. Most of my professional career has been devoted to understanding the motions of stars within galaxies, and motions of galaxies in the universe. The articles which follow discuss our Galaxy, some nearby galaxies, and a few others that were, and remain, of particular interest.

"The Past Decade Meets the Next Decade" appeared in the weekly *Science*—in a 1980 issue devoted to the accomplishments of the past decade and the promises of the coming decade.

"Structure and Evolution of the Galactic System" was a report from a 1960 international summer school for young astronomers held in Nyenrode Castle, Breukelen, The Netherlands. It was published in *Physics Today*. The course was a significant event in my professional career; the lectures brought me up-to-date on astronomical discoveries I had overlooked in the 1950s while getting my Ph.D. degree and raising four children with my husband, Bob. A paragraph, excluded from the reprint here, follows:

There were 65 students from 16 countries; 44 lectures by 21 speakers; a visit to the Leiden Observatory (bus) and Dwingeloo Radio Observatory (boat across the former Zuyderzee, now Ysselmeer); receptions by the mayor of Breukelen (another nearby castle), the city of Amsterdam (canal boat to the Municipal Museum) and the Board of Governors of NUFFIC at Muiderslot Castle (canal boat up the Vecht). Musical performances included a Chopin concert by Mr. George van Renesse, one of Holland's foremost pianists (in the lecture hall with concert grand, oriental rug, and candlelight), a lute folk singer at Muiderslot (in seven languages), and a concert by 12 harpists (which caused Westerhout to remark that the probability of so many astronomers and so many harpists together at one time was absolutely zero).

"Walking Through the Super(nova) Market" was an after dinner address given in Washington, 1977, at the annual meeting of The Association of Variable Star Observers, an international group of dedicated amateur astronomers who devote the dark hours to observing stars that vary in brightness. It has never before been published, and it is one of the very few early talks for which my notes have survived.

"Dynamics of the Andromeda Nebula" is a review from *Scientific American*, 1973, which includes the observations and analysis of the orbital velocities of stars and gas within our nearest large neighboring galaxy. This study started the long path to the discovery of flat rotation curves and dark matter surrounding virtually every galaxy.

"The Peculiar Galaxy NGC 1275" (1976), "NGC 3067" (1981), "S0 Galaxies with Polar Rings" (1982, 1983), and "A Pair of Noninteracting Spiral Galaxies" (1983) each appeared in the *Yearbook of the Carnegie Institution of Washington* in the indicated year. For many years, staff members wrote brief accounts of current research, most coauthored with colleagues. The study of polar ring galaxies was led by François Schweizer.

"UGC 2885, the Largest Known Spiral Galaxy" appeared in *The Astronomical Zoo* column of *Mercury*, 1980, the magazine of The Astronomical Society of the Pacific. This remarkable organization encourages interactions among its diverse membership: teachers, amateur astronomers, professional astronomers, and interested people from all walks of life.

"Some Surprises in M31 and M33" appeared in *The Outer Galaxy*, (edited by L. Blitz and F.J. Lockman, Springer–Verlag). It was a talk given in 1987 at the University of Maryland at a symposium in honor of Dr. Frank J. Kerr, a professor of astronomy and long-time friend. Part of the fun and joy of doing extragalactic astronomy in northwest Washington DC has been having Frank Kerr and his colleagues as neighbors.

"NGC 4550: A Two-Way Galaxy" describes the recent discovery of an amazing galaxy, whose characteristics have forced us to enlarge our ideas about galaxies and to alter procedures for analyzing their motions. It appeared in 1993 in *Mercury* magazine.

The Past Decade Meets the Next Decade

1980

Discoveries in astronomy over the past 20 years have challenged our perception of the universe. Attempts to understand these discoveries should direct the course of astronomy for years to come. Rarely in the history of science has there been an equivalent period in which the boundaries of our understanding have been expanded so dramatically.

Until the middle of this century, astronomical knowledge came from observations made in the optical region of the electromagnetic spectrum, that region transmitted by the earth's atmosphere and detected with the eye. This radiation has a frequency of 10^{15} cycles per second and a wavelength of 10^{-7} meter, and is characteristic of thermal radiation from stars like our sun with surface temperatures near 6000 °K. The universe that astronomers knew in the middle of this century was a majestic, slowly evolving stellar universe. But advancing technology has produced instruments which now detect radiation from the radio and infrared spectral regions, which are beyond the visible red end of the spectrum but are transmitted by the earth's atmosphere. In addition, advanced detectors fly above the earth's atmosphere and observe in the x-ray, γ-ray, ultraviolet, and far-infrared spectral regions, where the atmosphere is opaque. The observable spectrum has been enlarged to cover the wavelength range from 10^{-14} to 100 meters. Astronomers now know that we live in a veritable zoo where x-rays are emitted by objects as diverse as quasars, diffuse intergalactic gas, coronae about cool stars, matter accreting onto compact objects (perhaps black holes), and pulsating stellar remnants from supernova explosions; where γ-rays are a direct probe of cosmic nuclear processes and are produced from the interaction of cosmic rays with the interstellar matter and in supernovae, the sun, and compact galactic objects, sometimes in transient events lasting no more than 10 seconds; where nuclei of galaxies and some quasars radiate more in the infrared than they do in the visible; where sites of newly born stars are marked by infrared emission from dense molecular clouds; and where the vast regions between the stars in our galaxy contain complex organic chemical compounds, compounds that are fundamental constituents of living things on the earth. An exotic menu of astronomical sources is now routinely available for study, and the universe is known to be immeasurably richer, more varied, and more violent than would have been dreamed even 20 years ago. Glimpses of this variety are described below, and a few obvious paths for astronomical investigation in the 1980s are identified.

Cosmology

At its broadest, astronomy is the study of the universe. By observations made here and now, astronomers attempt to deduce the early history of the universe and uncover factors that have determined its evolution to the present. Most astronomers accept as a model a universe that has expanded and cooled from an initially hot, dense state. The primeval fireball radiation arose from the Big Bang, the inception of the expansion. This radiation has been expanding and cooling during 10- to 20-billion years since the Big Bang; its present temperature is 3 °K. After George Gamow's cosmological studies, the existence of this background microwave radiation was predicted in the late 1940s by Alpher and Herman [1], but it was not detected until 1965 by Penzias and Wilson [2]. Current research in cosmology is dominated by the impact of this discovery.

Thermal radiation of 3 °K (blackbody radiation to the physicist) has a characteristic spectrum with its peak radiation at a wavelength of 1 millimeter in the microwave spectral region, a region in which the earth's atmosphere is radiating with a temperature near 300 °K. Hence, accurate measurements of the spectral energy distribution of the background radiation can only be made above the earth's atmosphere. An outstanding achievement during the 1970s was the verification of the blackbody nature of the microwave radiation from instrumentation flown in a balloon [3]. However, tantalizing small departures from blackbody radiation may have been detected [4]. Such deviations are important because they make it possible for us to differentiate between events arising during the early thermal history of the universe and more recent effects, such as the radiation of warm dust from intervening galaxies along the line of sight.

The degree to which the radiation is isotropic—that is, the same in all directions—is a test of Big Bang cosmology; a measure of the initial shear, rotation, and inhomogeneities of the early universe; and acts as a speed-ometer for the motion of our galaxy. We are immersed in a sea of photons, photons which outnumber nucleons by a factor of 10^8, photons with ener-gies equivalent to a temperature of 3 °K. As our galaxy moves through these photons, we will see a hotter temperature in the direction of our motion and a correspondingly cooler temperature in the direction from which we have come. Within the past few years, a surprisingly large anisot-ropy in the background radiation has apparently been detected [5]. With respect to the background radiation, the galaxy and the local group of galaxies have a velocity of about 400 kilometers per second toward the Virgo supercluster of galaxies. If confirmed, this motion would imply that the combined mass of galaxies in the Virgo supercluster is sufficient to slow slightly the expansion of the universe in our vicinity.

Because the microwave background radiation is presently our most ac-cessible probe of the early universe, crucial experiments will be designed in the 1980s to improve the accuracy of the measured spectrum and to place more stringent limits on the large-scale and small-scale isotropy. Only such observations can establish for certain that the radiation is a relic of the Big Bang rather than the superposition of numerous point sources. Such measurements will be made by COBE, the Cosmic Background Ex-plorer, an orbiting NASA instrument which will observe the background radiation in the millimeter and submillimeter range. We should enter the 1990s more knowledgeable about the details of this fossil radiation.

Efforts to map the expansion of the universe in the vicinity of our gal-axy have proved to be unexpectedly difficult. Knowledge of the distances and velocities is required for a large number of galaxies; distance determi-nations especially are fraught with complex selection and systematic ef-fects. Despite heroic efforts on the parts of numerous astronomers, the value of the velocity of expansion, the Hubble constant H, is probably uncertain by a factor of 2 [6]. The value of H is a scale factor in cosmol-ogy. It affects calculated luminosities, sizes, and densities of extragalactic objects and measures the age of the universe in the simplest cosmologies. Observations which can produce a value of H accurate to 10% or 20% are required if we are to know the age and the scale of our universe; this major observational effort will continue during the 1980s.

With the launch of the Space Telescope in the 1980s, astronomers will have a large, high-resolution optical telescope outside of the earth's atmo-sphere. With this superb instrument it will be possible to observe indi-vidual stars, clusters, novae, and ionized gas clouds in external galaxies to greater distances than were ever before possible. All of these objects serve as standard candles for determining distances to individual galaxies. Observations made with the Space Telescope, coupled with detailed ground-based studies carried out with the use of both classical and novel techniques, should resolve the present controversy concerning the value of H.

Is the universe open or closed? Is the mean mass density low enough so that the universe will continue to expand forever, or is the density sufficiently high that gravity can slow the expansion and ultimately halt it so that it reverses into a contraction? Although presently available evidence favors continued expansion, we really do not now have a convincing answer [7]. We do understand that tests devised earlier to tell if the universe is open or closed often tell instead about galaxy evolution. Some galaxies may have been brighter in the past, a past which included more active star formation [8]. Large galaxies in clusters may have been fainter in the past, before they brightened by cannibalizing stars in the outlying regions of neighboring galaxies or by merging completely with their smaller neighbors [9]. Until we understand the luminosity history of galaxies, we cannot map distances in the early universe by measuring luminosities. Alternative approaches which attempt to enumerate directly the mass of the universe fail. The curious reasons for this failure are discussed below.

Quasars

The discovery of quasi-stellar objects (quasars) in the early 1960s had implications far beyond the astronomy and physics of these particular objects. Astronomers learned that a major constituent of the universe had until then been undiscovered; there is little doubt that other major components of our universe remain unknown today.

Quasars were identified initially by their intense radio emission [10]. Optical studies showed pointlike stellar sources whose spectra indicated enormous redshifts of strong emission lines [11]. In an expanding universe, spectral lines shifted redward arise from a velocity of recession; quasars with $z = 3.5$ (that is, wavelengths of lines shifted from the laboratory positions by a factor of 3.5) are the most distant objects known in the universe. Today, most astronomers agree that quasars are abnormal nuclei of very distant galaxies, radiating with enormously high luminosities. Some quasars have been found to reside in faint clusters of normal galaxies; the quasar and the cluster galaxies have the same redshift and hence are at the same distance. Some quasars have faint surrounding material, identified as the normal galaxy disk [12].

In one extraordinary case, two quasars extremely close together on the sky have virtually identical optical spectra and redshifts [13]; this finding may mean that we are observing two images of a single object. An intervening galaxy along the line of sight deflects the radiation from the background quasar, thereby acting as a gravitational lens and forming two (or perhaps three) images. If this model is correct, a long-predicted phenomenon will have been discovered. Moreover, there will be decisive evidence that the quasar is more distant than the intervening galaxy.

Quasars are observed to vary in brightness on incredibly short time scales [14]. At optical and infrared wavelengths, some magnitudes vary on a time scale of a day; some polarizations vary on a time scale of hours. The enormous x-ray intensity of quasars (10^{14} solar luminosities per second) can change on a time scale of hours. The short light travel times involved indicate that the central energy source is tiny, having only solar-system dimensions! From the plethora of models for quasars and active galaxy nuclei, accretion onto massive black holes of between 10^6 and 10^{10} solar masses now appears most satisfactory. Stars approaching near the black hole are torn apart and swallowed, releasing enormous amounts of gravitational energy. Regardless of whether this specific model proves to be correct, most astronomers agree that there are no compelling reasons to doubt that the observed redshifts indicate enormous distances or to believe that "new physics" is required to understand quasars. Still, the puzzle posed by their energetics is one of the most challenging in contemporary astronomy.

A continuing enigma arises from the observations of quasars which are double radio sources. For some of these, it appears that the radio sources are separating with superluminal (faster than the speed of light) velocities, if the quasars are placed at the cosmological distances inferred from their redshifts. At the present time, such observations are not interpreted as a threat to the current understanding of quasars. Rather, the radio signals are assumed to be generated or scattered or reflected in a stationary medium surrounding the active central object [15]. Such sources must be related to the jets seen on optical photographs of some radio galaxies. Continued interferometric observations during the 1980s should delineate their properties still further.

The numerous absorption lines seen in the spectra of quasars apparently have a variety of origins [16]. Some arise in the quasar, some in the surrounding galaxy, some in intervening galaxies, and some in intergalactic clouds along the line of sight. Such clouds will provide a unique probe of gas densities and abundances at earlier epochs in the universe. It is hoped that studies of these ubiquitous clouds will tell us the chemical and evolutionary history of both the clouds and the primeval galaxies which formed from them.

Quasars were more numerous and more luminous in the past [17]. The numbers of detected quasars decline at $z = 3$, and the most distant quasar has $z = 3.5$. Quasars thus are useful as a tool for identifying distant galaxies with z up to 3.5. What should we look for to identify still more distant galaxies, galaxies with z near 10? Will they be diffuse, very extended, have very low surface brightness regions, be red from their large red shifts, or will they be abnormally blue, signaling an enormous rate of star formation in those early days? And what characteristics identify the epoch of galaxy formation after the Big Bang? These are questions which astronomers are now trying to answer, often with the use of novel observing techniques, and which will be critical areas of study in the 1980s.

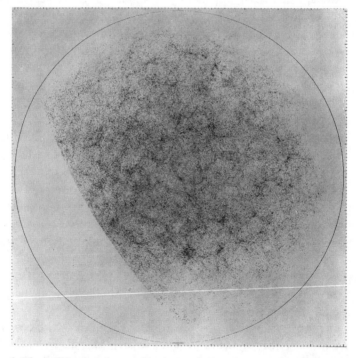

Figure 1. The million brightest galaxies as they appear on the northern sky, from counts by Shane and Wirtanen at Lick Observatory [46], newly reduced [18]. The north galactic pole is at the center, the galactic equator is at the edge, and galactic latitude is a linear function of radius. Note the striking lacework pattern and the conspicuous voids. [Courtesy of P. J. E. Peebles, from M. Seldner, B. L. Siebners, E. J. Groth, and P. J. E. Peebles, Astron. J. 82, 249 (1977).]

Distribution of Mass in the Universe

The distribution of visible matter in the universe is hierarchical, progressing from galaxies to clusters of galaxies to clusters of clusters. Counts of the million brightest galaxies (Fig. 1) reveal a lacy network of extended linear arrangements and great voids, with a surprising lack of isolated field galaxies [18]. Most galaxies reside in small groups or clusters (Fig. 2) which in turn clump to form superclusters. It was during the 1970s that astronomers realized that galaxies, after formation, are not the isolated island universes which Hubble envisioned earlier. Rather, they interact with

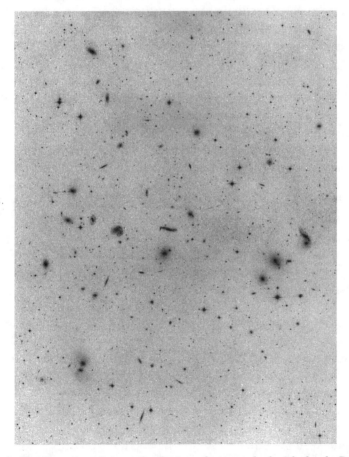

Figure 2. The Hercules cluster of galaxies, photographed with the du Pont 2.5-meter telescope at Las Campanas, Chile. Note the incredible variety of forms of galaxies contained within the cluster. (Courtesy of A. Dressler, Hale Observatories.)

their environment and with each other in complex ways. Elliptical galaxies are preferentially found in the higher density regions of clusters; spiral galaxies are most often found in the low-density outer cluster regions or in isolation [19]. Within clusters, central galaxies grow massive at the expense of halo stars from near neighbors. Galaxies in collision or galaxies tidally distorting each other produce the pathological forms often observed on the sky [20]. The galaxy environment appears fundamental in determining the morphology of the galaxy.

Additional factors which cause a galaxy to evolve as a spiral or as an elliptical remain unknown, although the local mass density and the local gas content must both be important. The gas content particularly is significant in directing the evolution of a galaxy, for from it new stars are formed. We do not yet know if galaxies preferentially accrete gas or lose gas to the intercluster spaces during their lifetimes; perhaps they do each at a different stage in their evolution. For massive galaxies at rest in the centers of large clusters, there is observational evidence of infalling hydrogen clouds. For galaxies moving at large velocities through the intercluster medium, it is expected that gas will be stripped from the galaxy as a result of the ram pressure of the intercluster gas. Within galaxies, high-velocity stellar winds, stirring by supernova explosions, shedding of cocoons by newly formed stars, and evaporation by a hot intergalactic medium will alter the balance between gas, dust, and stars. The delicate interplay of these processes will establish the presence or absence of a significant gaseous component [21].

Stars within galaxies evolve; galaxies within clusters evolve; and clusters of galaxies evolve. The present epoch may be called the epoch of cluster evolution. A cluster such as the Virgo supercluster (of which our galaxy is a suburban member) is in an early evolutionary stage, with an irregular shape, a large fraction of spirals, small random motions among the galaxies, low intercluster gas temperature, and low x-ray luminosity clumped around individual galaxies. As clusters evolve, the cluster shape becomes more regular, the spread in velocities among the galaxies increases, the central gas density increases (perhaps as a result of the stripping of gas from galaxies which pass through the core of the cluster), the central gravitational potential grows, and a supergiant elliptical galaxy may form at the center [22]. Astronomers were startled to learn that the hot intracluster gas, identified by its x-ray emission, is not the pristine hydrogen and helium formed shortly after the Big Bang and left over after galaxy formation, but is rich in heavy elements such as iron [23]. This is a certain sign that this gas has been synthesized in stellar interiors and returned by way of supernova explosions to the intergalactic spaces. Programs for the 1980s will attempt to learn how the evolution of galaxies within clusters has affected the evolution of clusters, and vice versa. Astronomy in the last 50 years has grown from the study of stars to the study of galaxies; the 1980s should be a time devoted to the study of clusters of galaxies.

Only during the last decade have astronomers acknowledged that much of the mass of the universe must be invisible, although the controversial evidence had been accumulating for a long time. Almost 50 years ago, Smith [24] and Zwicky [25] made an amazing observation; individual motions of galaxies in a cluster are so large that the gravitational attraction of all the cluster galaxies is not sufficient to bind the cluster. Galaxy clusters should thus be dissolving, although they apparently are not. This suggests that an unseen component of matter is present to bind the clusters. Very recent work has strengthened this conclusion; the dynamics of individual galaxies, double galaxies, groups, and clusters all point to this unobserved but ubiquitous mass component [26]. As much as 90% of the mass of the universe may be presently unseen. Its luminosity per unit mass must be considerably below that of the usual stellar matter. Astronomers are fond of saying that such mass could be in the form of bricks, or baseballs, or Jupiters, or comets, or mini-black holes. At present, its signature is known only by its gravitational interaction, but continued studies in all regions of the electromagnetic spectrum should help delineate its properties. The presence of such mass in quantities sufficient to bind the clusters could still be insufficient to close the universe.

At least some fraction of the nonluminous matter in the universe is located in the outer parts of individual spiral galaxies. Astronomers have long known that the stars and gas in a spiral galaxy are orbiting about the center of the galaxy. It was anticipated that the orbital velocities of stars would decrease with increasing distance from the center of the galaxy, just as the velocities of planets in the solar system decrease with increasing distance from the sun. Decreasing velocities arise as the gravitational response to a massive central body, that is, the sun in the case of the solar system. However, recent spectroscopic studies of spiral galaxies [27] show without doubt that the velocities of gas and stars remain high at large distances from the center (Fig. 3). This signifies that the mass in a galaxy is not as centrally condensed as it is in a solar system. In ordinary spirals, mass must be distributed far beyond the optical image, probably in massive dark halos.

A hot gaseous corona surrounding our galaxy may recently have been detected [28] on the basis of its characteristic ultraviolet lines seen in spectra obtained by the orbiting International Ultraviolet Explorer. Preliminary analysis indicates a halo of highly ionized atoms of carbon and silicon (suggesting a temperature of $\sim10^5$ °K) as well as less highly ionized spe-

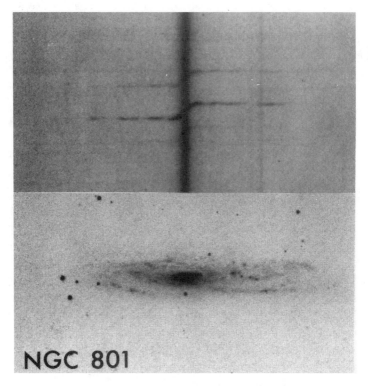

NGC 801

Figure 3. *(Bottom) NGC 801, a spiral galaxy seen close to edge on, from a Kitt Peak National Observatory plate taken with the 4-meter telescope. (Top) Emission lines along the major axis of NGC 801 coming from excited gas in the galactic disk (horizontal); the stronger is due to hydrogen. The vertical stripe arises from stars in the nucleus. As stars and gas orbit in NGC 801, gas on the SE (left) approaches the observer, so emission lines are shifted to short wavelength. Gas on the NW (right) is receding from the observer, so lines are shifted to long wavelengths (up on print). The high velocity gradient near the nucleus produces the highly inclined lines seen there. The gravitational influence of low luminosity mass at large nuclear distances produces constant rotational velocities observed at large nuclear distances. (Photo courtesy of B. Carney, W. K. Ford, Jr., and V. C. Rubin.)*

cies. The discovery of a hot halo about our galaxy strengthens the evidence that some of the quasar absorption lines originate in halos of foreground galaxies along the line of sight to the quasar.

Elliptical galaxies also show unexpected dynamical properties. These galaxies appear as spheroidal systems exhibiting various degrees of flattening; astronomers had assumed that the highly flattened configuration arose as the result of rapid rotation. Recent dynamical studies have shown that this is not so [29]. Many flattened elliptical galaxies are rotating only slowly, too slowly for rotation to be the cause of their flattening. Elliptical galaxies may have three unequal axes, with stable stellar orbits aligned in a triaxial shape. Classical stellar dynamics is merging with modern cosmology as efforts are being made to understand the dynamics of both elliptical and spiral galaxies.

Our Galaxy

Surprisingly, progress in deciphering the structure of our own galaxy has not kept pace with extragalactic achievements. We know that we live in a spiral galaxy although its detailed morphology and dimensions remain a mystery. We do not know how far our sun is from the center, nor do we know our rotational velocity about the center with an accuracy sufficient to determine the galactic scale to within 20%. Astronomers now understand spiral arms as a wave phenomenon [30], but the theory is more successful in the general than in the specific. Initial progress in deducing the detailed structure of the distant nucleus of our galaxy has come from very-long-baseline interferometry in the radio spectrum and from observations of ionized neon emission in the infrared [31]. Within a small region at the center of our galaxy, there is a bewildering asymmetrical configuration of small sources of various sizes and velocities. The mass is high and the energy output is high, especially in the infrared.

Our galaxy is not unique in its nuclear energetics. Nuclei of galaxies are sources of enormous energy, radiating in the x-ray, optical, infrared, and radio spectral regions. Quasars, Seyfert galaxies, BL Lac objects and even normal galaxies produce great amounts of energy. All of these objects may contain a central massive object, perhaps a black hole, which devours approaching stars and gas with the release of enormous gravitational energy. Alternatively, perhaps conventional astrophysical phenomena can account for the observed effects. Understanding the source of the central energy "machine" and its effect on star formation and on the evolution of galaxies is an important goal of astronomy in the 1980's.

During the coming decade, the variety of approaches available for the detailed study of our galaxy should produce a more coherent picture of the stellar system in which we live. From the Space Telescope, studies of faint halo stars, distant globular clusters, and outlying satellite galaxies should delineate the extent of the system and its chemical evolution as a function of age and of position; it is hoped that radio and millimeter observations of molecular clouds will identify regions of star formation and help us to learn their dynamics. Sophisticated theoretical models will address the stability of disk systems. From these investigations astronomers should learn what role warps in the outer disk play in this stability. Perhaps the Magellanic Stream [32], that extensive band of neutral hydrogen reaching from our galaxy to the nearest satellite neighbors, will be understood in this context. From x-ray observations of nearby galaxies, we should learn about energetic nuclear events. Our nearest spiral neighbor, M31, now has 17 sources of x-ray radiation identified [33] in its nucleus; their nature is not yet understood. We may ultimately be able to piece together a detailed picture of our own galaxy from studies both internal and external.

Stellar Evolution

Crucial to our understanding of star formation is a knowledge of the interstellar gas and dust from which new generations of stars are born. During the 1970s, astronomers learned that the interstellar medium in our galaxy has both hot and cold components [34]. Some gas is at a temperature of 1,000,000 °K; some is in the form of dense molecular clouds with temperatures near 10 °K. New instruments in the millimeter and infrared spectral regions have permitted astronomers to probe the cold molecular clouds, the birthplaces of stars. The cocoons of gas and dust contract under their own gravity, warming as they collapse, until their core temperatures become high enough to initiate nuclear fusion; a protostar or cluster of protostars is born.

There is some evidence that supernova explosions from dying stars compress the clouds [35], initiating a new generation of star births. Both spatially and temporally, star formation appears sequential [36], with new generations arising at the outer edges of older associations. The roles played by magnetic fields, rotation, and turbulence and the factors that control single or multiple births are not known. Like the sun, some stars retain a disk or a complex pattern of debris after their formation. Such stars are candidates for associated solar systems.

Most of a star's lifetime is spent in a stable phase, during which the star draws upon its nuclear energy sources to maintain its energy supply; details of its evolution depend, among other factors, upon its mass and composition. It has become increasingly clear, however, that stars of almost all masses lose significant amounts of mass during their evolution, either in active winds or explosive events [37]. From observations in the ultraviolet, optical, and infrared, a crude picture of the processes involved has been drawn. Interestingly, it now appears likely that the interstellar dust originates in the stellar ejecta. Knowledge of stellar interiors, stellar mixing, and ejection mechanisms will have to precede an understanding of the cycling, nucleosynthesis, and recycling of gas through stars.

An energy crisis occurs during the late stages of stellar evolution, when a star has exhausted its nuclear fuel or when further heating by contraction is not possible. We can now identify white dwarfs, neutron stars, and probably black holes as the end products of stellar evolution. Observations of Sirius B established the existence of white dwarfs, and the identification of the pulsar in the Crab Nebula established the existence of neutron stars. Unfortunately, there is as yet no corresponding conclusive evidence for the existence of black holes, although candidates continue to emerge. One clue to their existence is the x-rays emitted by matter accreting onto massive compact objects. Exotic objects recently identified, x-ray bursters and γ-ray bursters, are both related to such circumstances, although little is yet certain about the mechanisms involved. In view of the intense observations and theoretical studies concerning the reality of black holes, it is interesting to recall the discovery of the first white dwarf, Sirius B. In 1844, Bessel noted the irregular motion of Sirius and concluded that it was a double star. In 1862, the firm of Alvan Clark, the American telescope maker, completed an 18-inch lens ordered by the University of Mississippi. (Because of the Civil War, the lens was never delivered and ultimately ended up at the Dearborn Observatory of Northwestern University.) In testing the lens apart from a telescope, Alvan Clark, Jr., detected the faint companion [38]. However, not until 50 years later was a spectrum obtained, and not until 10 years after that was the astrophysical significance of a small but massive star recognized. The pace of science is faster today, and we hope that it will not be 80 years until direct evidence for black holes is established.

Neutron stars have masses like that of the sun, but radii 10^5 smaller and densities 10^{14} larger. The first neutron star was discovered [39] as a radio pulsar; the pulsed radiation signature arises from the presence of an intense magnetic field. Presently about 300 pulsars within our galaxy are known. Some of these are in close double systems; x-ray emission in these systems is understood as the accretion of mass onto the neutron star. Changes in the observed period of pulsation give insights into exceedingly complex physical conditions. With their solid lattice surfaces and superfluid cores, pulsars are laboratories for physical conditions not attainable on the earth.

A binary pulsar whose orbital period has been monitored for 5 years [40] has produced the first convincing indirect evidence of the existence of gravitational waves, long predicted as a result of the general theory of relativity. The small but continual shortening of the orbital motion of the two objects is attributed to the energy dissipated by the radiation of gravitational waves. Experiments to detect directly the gravitational radiation from more intense sources (collapsing cores of supernovae, for example) are extremely difficult but are presently underway [41].

The Sun

Solar physics during the 1970s has been evolving from an intensive study of our closest star to a broader inquiry into the behavior of ionized gas in gravitational and electromagnetic fields in stars. The sun acts as an almost "hands on" laboratory which can provide insight into phenomena such as mass loss, hot coronae, activity cycles, and heating. All of these processes are known to occur in a wide variety of stars.

The discovery of an unexpectedly low neutrino flux from the sun has forced a broad reevaluation of ideas concerning the structure and dynamics of the solar interior. The probable detection of neutrinos [42] at the level of 2.2 ± 0.4 solar neutrino units still remains too low a value by several times in comparison with the value predicted on the basis of recent laboratory and theoretical results. A reexamination of stellar models, abundances, and nuclear parameters, and of questions such as whether the sun has a rapidly rotating core or whether the interior undergoes episodes of mixing, will be imperative if we are to understand the solar interior. We are forced to conclude that the rate of burning in the solar interior is not as well understood as was once thought.

Understanding the nature of both large- and small-scale structures on the sun and their circulation is essential. Such motions are intimately connected with solar activity and possibly with long-term fluctuations of the sun, both of which may directly affect terrestrial climate. The surprising discovery that the solar atmosphere has radial wave motions with a 5-minute period [43] has led to major theoretical studies of the generation, propagation, and damping of sound, gravity, and magnetic waves. The 5-minute oscillations are now being used to probe part of the solar interior in a manner analogous to seismological probing of the core structure of the earth.

George Hale's early investigations of solar magnetic fields made it possible to distinguish between strong magnetic fields in sunspots and a weak, 1-gauss general magnetic field of the sun. During the last decade, sophisticated observing techniques have revolutionized our concept of the solar magnetic field structure [44]. In regions where a solar magnetic field occurs, its strength is very high (1500 gauss). The 1-gauss general magnetic field observed earlier was the result of the small size of these high-field

magnetic elements, which were averaged over by the low-resolution mag-
netometers. Observationally, their true size is unknown; they are too small
for resolution by even the best ground-based magnetograms. They may
have the size of solar "filigree," the smallest structures that have been seen
on the sun (< 0.3 arcseconds), which appear to be cospatial with the magnetic elements. Theoretical attempts to explain the origin and stabilities of
these magnetic flux tubes are under way.

The study of solar magnetism at a level of detail and quality never before possible will be a prime objective of the Solar Optical Telescope on
the Space Shuttle. Major advances in our knowledge of solar (and, by
implication, stellar) magnetic fields and their relation to solar activity,
coronal structure, and heating are anticipated in the 1980s.

Several outstanding discoveries of solar physics during the past decade
are related to the corona [45]. The solar corona contains active regions
with strong magnetic fields and closed (loop) configurations; quiet regions
with weak fields, apparently closed on large scales; and coronal holes associated with weak magnetic fields having a diverging open configuration. Coronal bright points are small regions of intense x-ray and extreme
ultraviolet emission associated with newly emerging magnetic flux into
the corona. Coronal holes are the source of recurrent high-speed solar wind
streams (1000 kilometers per second). Coronal transients are frequent mass-
ejection events of large energy, important in restructuring the outer corona
and in the flare process. All of these phenomena must be understood if we
are to build a meaningful foundation for stellar wind theory. During the
1980s, satellites carrying x-ray telescopes and magnetographs with high-
resolution imaging capability should provide data needed by theoreticians
to address these problems not only for the sun but also for their astrophysical and geophysical implications.

Although the study of the stars and the sun leads ultimately to the study
of the origin of the solar system, at present only a handful of astronomers
and geophysicists are at work on these problems. Studies of abundances in
meteorites, the debris from the formation of the planets, of planets and
small particles in the solar system, of nuclear chronology, of stars just
forming and stars with associated disks, all should increase our understanding of what it takes to make a solar system. Searches for planets
about other stars, carried out with a variety of techniques including imaging them directly or noting their effect on the motions of the parent star,
are likely to be undertaken during the 1980s. If we can detect other solar
systems, then we may be able ultimately to detect other civilizations. And
the detection of and eventual communication with others in the universe is
truly a goal for the future.

Conclusions

We live in a fascinatingly beautiful universe. For thousands of years, people have been looking at the sky and enumerating the objects seen; for hundreds of years astronomers have been using telescopes to discover fainter components. Only during the last 30 years have astronomers and physicists been able to look at the sky in wavelength bands other than that of visible light. Such views have shown us entirely new classes of objects and phenomena never imagined.

But astronomers during the past few decades have done more than discover and catalog. Fundamental insights into the complex interplay of all facets of the universe have emerged. Stars form from primordial gas, synthesize heavy elements, and return the enriched gas to the interstellar spaces as material for new generations of stars. The dynamics of spiral structure affect the star formation process in the galactic disk; the rates of star formation influence galaxy morphology. The particular cluster environment influences the morphology of galaxies, whether spiral or elliptical; cluster characteristics reflect galaxy evolution. The molecules in our bodies are formed from atoms synthesized in some star and deposited, probably explosively, into the primitive solar nebula. In the coming decades, it seems likely that astronomers will discover still more bizarre objects and will study their properties with innovative instrumentation on the ground and in space. During the 1980s, the inaugural flight of the Space Telescope is expected to show us wonders we can now only dimly perceive. From all of these studies we should be able to determine more about how the universe works. In *My First Summer in the Sierra*, John Muir wrote [47], "When we try to pick out anything by itself, we find it hitched to everything else in the universe." The thrust of astronomical investigation in the next decade will be to elucidate some of these connections.

References and Notes

1. R.A. Alpher and R. Herman, Nature (London) **162**, 774 (1948).
2. A.A. Penzias and R.W. Wilson, Astrophys. J. **142**, 419 (1965).
3. D.P. Woody, J.C. Mather, N.S. Nishioka, P.L. Richards, Phys. Rev. Lett. **34**, 1036 (1975).
4. D.P. Woody and P.L. Richards, *ibid.* **42**, 925 (1979).
5. G.F. Smoot, M.V. Gorenstein, R.A. Muller, *ibid.* **39**, 898 (1977); E.S. Cheng, P.R. Saulson, D.T. Wilkinson, B.E. Corey, Astrophys. J. Lett. **232**, L139 (1979).
6. A. Sandage and G.A. Tammann, Astrophys. J. **210**, 7 (1976); G. de Vaucouleurs and G. Bollinger, *ibid.* **233**, 433 (1979); P.J.E. Peebles, Comments Astrophys. **6**, 197 (1978); R.B. Tully and J.R. Fisher, Astron. Astrophys. **54**, 661 (1977); M. Aaronson, J. Huchra, J. Mould, Astrophys. J. **229**, 1 (1979).

7. J.R. Gott III, J.E. Gunn, D.N. Schramm, B.M. Tinsley, Sci. Am. **234**, 62 (March 1976).

8. B.M. Tinsley, Astrophys. J. Lett. **173**, L93 (1973).

9. J.P. Ostriker and S.D. Tremaine, *ibid.* **202**, L113 (1975); A. Toomre, in *The Evolution of Galaxies and Stellar Populations*, B.M. Tinsley and R.B. Larson, Eds. (Yale University Observatory, New Haven, Conn., 1977), p. 401.

10. D.O. Edge, J.R. Shakeshaft, W.B. McAdam, J.E. Baldwin, S. Archer, Mem. R. Astron. Soc. **68**, 37 (1960); C. Hazard, M.B. Mackey, A.J. Shimmins, Nature (London) **197**, 1037 (1963).

11. T.A. Matthews and A.R. Sandage, Astrophys. J. **138**, 30 (1963); M. Schmidt, Nature (London) **197**, 1040 (1963).

12. A. Stockton, Nature (London) **274**, 342 (1978); J. Kristian, Astrophys. J. Lett. **179**, L61 (1973); E.J. Wampler, L.B. Robinson, E.M. Burbidge, J.A. Baldwin, *ibid.* **198**, L49 (1975).

13. D. Walsh, R.F. Carswell, R.J. Weymann, Nature (London) **279**, 381 (1979); D.H. Roberts, P.E. Greenfield, B.F. Burke, Science **205**, 894 (1979).

14. H. Smith and D. Hoffleit, Publ. Astron. Soc. Pac. **73**, 292 (1961); M. Schmidt, Annu. Rev. Astron. Astrophys. **7**, 527 (1969); H. Tananbaum *et al.*, Astrophys. J. Lett. **234**, L9 (1979).

15. G.A. Seielstad, M.H. Cohen, R.P. Linfield, A.T. Moffet, J.D. Romrey, R.T. Schilizzi, D.B. Shafer, Astrophys. J. **229**, 53 (1979).

16. J.N. Bahcall and L. Spitzer, Astrophys. J. Lett. **156**, L63 (1969); A.F. Davidsen, G.F. Hartig, W.G. Fastie, Nature (London) **269**, 203 (1977); P.A. Strittmatter and R.E. Williams, Annu. Rev. Astron. Astrophys. **14**, 307 (1976).

17. R.F. Green and M. Schmidt, Astrophys. J. Lett. **220**, L1 (1978).

18. M. Seldner, B. Siebers, E.J. Groth, P.J.E. Peebles, Astron. J. **82**, 249 (1977).

19. A. Dressler, Astrophys. J., in press.

20. A. Toomre and J. Toomre, *ibid.* **178**, 623 (1972).

21. W.G. Mathews and J.C. Baker, *ibid.* **170**, 241 (1971); S.M. Faber and J.S. Gallagher, *ibid.* **204**, 365 (1976).

22. C. Jones, F. Mandel, J. Schwarz, W. Forman, S.S. Murray, F.R. Harnden, Jr., Astrophys. J. Lett. **234**, L21 (1979); J.P. Ostriker, in *The Evolution of Galaxies and Stellar Populations*, B.M. Tinsley and R.B. Larson, Eds. (Yale University Observatory, New Haven, Conn., 1977), p. 369.

23. P.J. Serlemitsos, E.A. Boldt, S.S. Holt, R. Ramaty, A.F. Brisken, Astrophys. J. Lett. **184**, L1 (1973).

24. S. Smith, Astrophys. J. **83**, 23 (1936).

25. F. Zwicky, *ibid.* **86**, 217 (1937).

26. S.M. Faber and J.S. Gallagher, Annu. Rev. Astron. Astrophys. **17**, 135 (1979).

27. M.S. Roberts and A.H. Rots, Astron. Astrophys. **26**, 483 (1973); A. Bosma, thesis, Rijks-universiteit te Groningen (1978); V.C. Rubin, W.K. Ford, Jr., N. Thonnard, Astrophys. J. Lett. **225**, L107 (1978).

28. B.D. Savage and K.S. de Boer, Astrophys. J. Lett. **230**, L77 (1979).

29. F. Bertola and M. Capaccioli, Astrophys. J. **200**, 439 (1975); G. Illingworth, Astrophys. J. Lett. **218**, L43 (1977).

30. C.C. Lin, J. Soc. Ind. Appl. Math. **14**, 876 (1966); and F.H. Shu, Astrophys. J. **140**, 646 (1964); W.W. Roberts, Jr., M.S. Roberts, F.H. Shu, *ibid.* **196**, 381 (1975); E.B. Jensen, K.M. Strom, S.E. Strom, *ibid.* **209**, 748 (1976).

31. K.Y. Lo, R.T. Schilizzi, M.H. Cohen, H.N. Ross, Astrophys. J. Lett. **202**, L63 (1975),

E.R. Wollman, T.R. Geballe, J.H. Lacy, C.H. Townes, D.M. Rank, *ibid.* **218**, L103 (1977).

32. D.S. Mathewson and M.N. Cleary, Astrophys. J. **190**, 291 (1974).

33. L. Van Speybroeck, A. Epstein, W. Forman, R. Giacconi, C. Jones, W. Liller, L. Smarr, Astrophys. J. Lett. **234**, L45 (1979).

34. G. Field, in *Confrontation of Cosmological Theories with Observational Data*, M.S. Longair, Ed. (International Astronomical Union Symposium 63, Reidel, Dordrecht, Netherlands, 1974), p. 13; P. Thaddeus, in *Star Formation*, T. de Jong and A. Maeder, Eds. (Reidel, Dordrecht, Netherlands, 1974), p. 37.

35. W. Herbst and G.E. Assousa, Astrophys. J. **213**, 473 (1977).

36. B.G. Elmegreen and C.J. Lada, *ibid.* **214**, 725 (1977).

37. J.P. Cassinelli, Annu. Rev. Astron. Astrophys. **17**, 275 (1979).

38. A.M. Clerke, *A Popular History of Astronomy During the Nineteenth Century* (A. and C. Black, Edinburgh, 1885), p. 52.

39. A. Hewish, S.J. Bell, J.D.H. Pilkington, P.F. Scott, S.A. Collins, Nature (London) **217**, 709 (1968).

40. J.H. Taylor, L.A. Fowler, P.M. McCulloch, *ibid.* **277**, 437 (1979).

41. J. Weber, Sci. Am. **224**, 22 (May 1971).

42. J.N. Bahcall and R. Davis, Jr., Science **191**, 264 (1976).

43. R.B. Leighton, R.W. Noyes, G.W. Simon, Astrophys. J. **135**, 474 (1962); R.F. Stein and J. Leibacher, Annu. Rev. Astron. Astrophys. **12**, 407 (1974).

44. E.N. Parker, Sci. Am. **233**, 42 (September 1975); *Cosmical Magnetic Fields* (Oxford Univ. Press, London, 1979).

45. G.L. Withbroe and R.W. Noyes, Annu. Rev. Astron. Astrophys. **15**, 363 (1977); G.S. Vaiana and R. Rosner, *ibid.* **16**, 393 (1978).

46. C.D. Shane and C.A. Wirtanen, Publ. Lick Obs. 22 (part 1) (1967).

47. J. Muir, *My First Summer in the Sierra* (Houghton Mifflin, Boston, 1916, reprinted in 1979), page 157.

48. I thank Drs. J. Beckers, B. Carney, J.S. Gallagher, R. Herman, J. Linsky, R.J. Rubin, and J. Thomas, and A. Rubin for valuable comments on the manuscript; L.W. McKenzie, U.S. Department of the Interior, Yosemite National Park, for locating the quote by J. Muir; and M. Coder for her capable typing. The National Academy of Sciences is presently sponsoring a study of astronomy for the 1980's, the Astronomy Survey Committee, Dr. G. Field, chair. I thank the working groups on Extragalactic Astronomy (S.M. Faber, chair), Galactic Astronomy (R.D. Gehrz, chair), Solar Astronomy (A.B.C. Walker, chair), and Related Areas of Science (J.E. Gunn, chair), and the members of the UVOIR (Ultraviolet, Optical, and Infrared) Panel, Dr. J.E. Wampler, chair, for their preliminary reports, all of which have been of great value in identifying astronomical highlights during the past and coming decade.

Structure and Evolution
of the Galactic System

1960

E ver since the beginning of this century, the Netherlands has been an important center for fundamental research on the structure and dynamics of the galactic system. Work by Kapteyn (1904) and Hertzsprung (1905) has led to methods which form the basis of much current research in this field. Recently, the first comprehensive explorations of the galaxy at 21-cm wavelength were made in Holland. To enable young astronomers to hear, first hand, the results of current investigations, an international summer course in science was held July/August 1960 at Nyenrode Castle, The Netherlands. The galactic system was the topic, and Jan Oort was one of the principal lecturers (Fig. 1).

Distribution of Gas in the Galaxy

Because interstellar absorbing clouds make optical studies difficult, the large-scale structure of the galaxy is best determined by observations of the 21-cm emission line of neutral hydrogen, the principal component of the interstellar gas. The following model results from such studies. The gas is strongly concentrated in a thin disc, with a radius of 15 kpc (1 kpc = 10^3 pc; 1 parsec = 3.26 light years = 3.08 10^{18} cm) and a thickness of 220 pc. In the central region, the disc is exceedingly flat, but in the outer regions the gas layer tips up in one direction, down in the other. The distribution of gas is not uniform. Regions of high density extend in long bands around the system, suggesting spiral arm structure. The arms are highly circular, but irregular, with sudden breaks, crosses, and confused regions. The sun, at 8.2 kpc from the center, is located on the inner edge of one of these arms.

Figure 1. Professor Jan Oort at the blackboard with students at the Nyenrode *1960 summer school. To Oort's left is Dr. Arcadio Poveda (Mexico), and to his right is Dr. Donald Morton (Canada). To the far right is Professor P. O. Lindblad (Sweden).*

The entire system is rotating, not as a solid disc, but with angular velocities which decrease with increasing distance from the center. The result of this differential rotation is to produce a systematic effect, a double sine wave, in the radial velocities of nearby stars (with respect to the sun), as a function of longitude in the galactic plane. The existence of such an effect, first explained by Oort (1927), is one of the best-established facts of modern stellar dynamics. It is the analysis of the Doppler shifts in the observed 21-cm line, as a function of longitude, which leads to the picture of the interstellar gas outlined above.

The region about the center of the galaxy exhibits many interesting phenomena. Within 0.4 kpc of the center, no differential rotation or expansion velocities have been observed; perhaps this region is rotating as a solid disc. At about 0.6 kpc, a ring with high rotational velocities is found, but the velocities decrease rapidly with increasing distance from the center. Outside of this ring, expansion velocities in the disc as high as 200 km/sec are observed. The gas density decreases from the center, but at 3 kpc increases to form a very regular arm. The entire arm is expanding away from the center with a velocity of 53 km/sec. Unlike the patchy irregular arms in the outer regions, this 3-kpc arm is uniform in density and velocity and can be traced a considerable distance around the galactic center.

There is some evidence, though no direct observations, for the existence of a gaseous spherical halo about the galaxy, with a diameter of 15 kpc. A density of 10^{-3} atoms/cc and a temperature of 10^6 °K (Spitzer) and a magnetic field of the order of 10^{-6} gauss follow from considerations of the background synchrotron radio emission of the galaxy and the magnetic field necessary to contain cosmic-ray particles. The most widely accepted theory assumes that the mass of gas expanding from the center region of the galaxy, 1 solar mass/year, is matter from the halo that streams down into the center of the galaxy. To retain the halo, Woltjer has postulated the existence of an intergalactic gas (density 10^{-4} atoms/cc, $T = 0.5 \times 10^6$ °K) outside the halo and between the galaxies which make up the local group of galaxies. On this model, it is the motion of our galaxy into the local group, and through the intergalactic medium, which has caused the distortion of the galactic plane.

Optical Studies of Galactic Structure

The large-scale distribution of stars in the galactic plane is more difficult to determine than the distribution of gas. Interstellar absorption complicates distance determinations and prohibits observations of stars at distances greater than a few kiloparsecs from the sun, except in a few windows. Much progress has been made as a result of Baade's (1944) observation that the central regions of external galaxies contain principally red giants, many RR Lyrae variables, but no gas or dust. In contrast, the spiral structure is delineated by the hot O and B stars, galactic clusters, and the presence of gas, dust, and emission regions. Baade introduced the term "populations" to differentiate the two distributions: population II for the nucleus and halo population; population I for the disc—O, B stars, gas and dust—population.

This concept of two populations has been applied to our galaxy to investigate the spiral structure and determine the distance from the sun to the center. Frequency counts of RR Lyrae variables as a function of increasing faintness, in the approximate direction of the center of the galaxy, led Baade to adopt 8.16 kpc as the distance of the sun from the galactic center.

The determination of the spiral structure has not been as successful. While studies of O and B stars by Morgan, Sharpless, and Osterbrock (1952) indicate that regions of ionized hydrogen are arranged in long strings, little over-all structure has been outlined. It has long been hoped that the O and B associations or Cepheid variables could be used for this purpose. To date, the results have been more promising than conclusive. The spiral structure of our galaxy would probably be unsuspected even today, if it were not for the observations of spiral structure in external galaxies.

The analysis of the motions of the stars, near the sun, in terms of the Oort theory of differential rotation, determines the values of Oort's constants, A and B:

$$A = \frac{1}{2}\left(\frac{V}{R} - \frac{dV}{dR}\right)_{\odot}$$
$$= +18.6 \pm 1.5 \text{ km / sec kpc} = +6.04 \times 10^{-16} \text{ sec}^{-1}, \tag{2.1}$$

$$B = -\frac{1}{2}\left(\frac{V}{R} + \frac{dV}{dR}\right)_{\odot}$$
$$= -7 \pm 1.5 \text{ km / sec kpc} = -2.3 \times 10^{-16} \text{ sec}^{-1}, \tag{2.2}$$

where V and R are the circular velocity and distance to the center respectively, evaluated at the sun ($\odot$). K_R, the R component of the gravitational field at the sun, may be evaluated in terms of A and B as:

$$\frac{K_R}{R} = -(A-B)^2; \quad \frac{\partial K_R}{\partial R} = (A-B)(3A+B). \tag{2.3}$$

From stellar velocities and star counts at right angles to the galactic plane, at the location of the sun, Schmidt has constructed a mass model of the galaxy and tabulated the gravitational potential and the R and z components of the gravitational field. At 0.2 kpc above the plane, the star density has fallen to 1/2 that at the sun; at 1.2 kpc it is down to 1/60, and it decreases as z^{-2} thereafter. At these heights above the galactic plane, the stars are halo population II. It is estimated that they represent 2% of the stars in the solar neighborhood. The period of revolution about the galaxy at the position of the sun is 233×10^6 years; for stars with $|z| < 1.5$ kpc, the vibrational period perpendicular to the plane is 66×10^6 years. On this model, the density of matter at the sun is of the order of 10×10^{-24} gm/cc $= 0.15$ solar mass/cubic pc.

Optical studies of the polarization of starlight (Hiltner, Hall, 1949) indicate that light only from reddened stars is polarized; for stars in low galactic latitudes the plane of polarization often is approximately parallel to the galactic plane. This was emphasized by Shajn (1958) who noted that many of the dark filaments in the galaxy are very elongated, generally in a direction parallel to the plane of polarization of starlight in the surrounding regions. The presence of elongated grains in interstellar space, and a magnetic field capable of aligning the grains, will explain these effects. Although the physical chemistry of the grains is still uncertain, Chandrasekhar and Fermi (1953) estimated that a magnetic field in the arms of the order of 10^{-5} gauss is consistent with the observations.

Evolution of the Galaxy

The giant step, from the discovery of the structure of the galaxy to the understanding of the evolution of the system, has not yet been taken. However, the evolution of individual stars and star clusters can be understood and indicates the general direction in which the galaxy must be evolving. Ages of stars, and clusters also, range from 10^6 to 10^{10} years, so star formation must be a continuing process. Comparison of star counts in gaseous regions with the mass of gas indicates that the rate of star formation is approximately proportional to the mass of gas. The bright O and B stars, usually associated with gas and dust, are the youngest stars; population I is the young population. Globular clusters, on the other hand, have been assumed to be devoid of gas, so presumably no star formation is proceeding in them; population II is the old population. Observations of dark specks in globular clusters (Roberts) interpreted as gas or even protostars, plus the presence of a few bright (young?) stars in globular clusters, may necessitate a change in this simple picture, however.

External galaxies, too, are classified into two groups: spiral galaxies which contain not only population II objects, but also gas, dust, O and B stars, and spiral arms; elliptical galaxies, with no gas, no O and B stars, no spiral arms. Unlike clusters, however, which may be presumed to have a single population of stars, galaxies are much less homogeneous. Thus there are elliptical galaxies with dark lanes, galaxies with O and B stars but no observable gas, irregular and pathological galaxies of all descriptions. G. R. Burbidge has suggested that elliptical galaxies form stars under conditions which remain uniform in time. In a spiral galaxy, conditions change with time; perhaps fourth-generation stars exist now in our galaxy.

The evolution of one structural feature of our galaxy, the spiral arms, may be examined in more detail. From observations of external galaxies it appears that spiral structure is a common phenomenon in some types of galaxies, yet differential rotation is also common. In our galaxy, a star 2 kpc from the sun (6.2 kpc from the center) moves faster than the sun, and would be 180° ahead of the sun in 3×10^8 years, a time only 1/30 the age of the galaxy. Hence the arms would be a very transient feature. If, following Oort, we adopt the attitude that new arms are not continuously formed, there are only several possibilities. The gas and stars in the arms may not move in circular orbits, but may stream in or out along the arms; or the arms may grow on one side, deplete on the other, and thus change position with time; or there may be a dynamical explanation, such as that offered by P. O. Lindblad.

Using the concept of dispersion orbits introduced by B. Lindblad, P. O. Lindblad has followed by computer the motion of 192 mass points in the Schmidt gravitational field. For particular initial density conditions, leading spiral arms will form, unwind due to the differential rotation, and finally trail. Continued rotation will wind the arms tighter, until an almost smeared-out density will result, close enough to the initial conditions to again start the formation of arms. Certainly, the study of the evolution of spiral arms is just beginning.

Most of the details of galactic structure are uncertain or open to revision. With no attempt at completeness, here is a list of some outstanding problems:

1. Spiral structure—why; how maintained against differential rotation; stellar or gas phenomenon; position of elliptical galaxies in evolutionary sequence?
2. Dynamics—are gas motions same as stellar motions; expansion or contraction in outer regions; replenishment of gas at center?
3. Interstellar dust—classical physical chemistry or quantum mechanics?
4. Galactic halo—does it exist; structure; polarization of radio signals, Faraday effect; mass due to type II novae?

This is the field, according to van de Hulst, in which everything over ten years old is classical. In June 1950, the University of Michigan dedicated a new telescope with a symposium on The Structure of the Galaxy. Spiral structure of our galaxy was then only a vague feeling induced by observations of external galaxies. Yet the unsuspected error in the distance scale made even this analogy difficult. Radio astronomy was nowhere even mentioned. Thus after ten years of exceptional progress it is strange to note that many of the same problems are still with us. Let us increase our time scale by a factor of ten. June 29, 1850, a paper was read before the Royal Society, by the Earl of Rosse, who five years earlier had discovered spiral structure. And in a sentence which could have been uttered in 1960, he said, "The sketches which accompany this paper are on a very small scale, but they are sufficient to convey a pretty accurate idea of the peculiarities of structure which have gradually become known to us: in many of the nebulae they are very remarkable, and seem even to indicate the presence of dynamical laws we may perhaps fancy to be almost within our grasp."

Walking Through the Super(nova) Market

1977

One evening, when I was contemplating as usual the celestial vault, whose aspect was so familiar to me, I saw, with inexpressible astonishment, near the zenith, in Cassiopeia, a radiant star of extraordinary magnitude. Struck with surprise, I could hardly believe my eyes ... —*Tycho Brahe*, November 1572

Tycho Brahe was luckier than most of us will be. He saw a supernova with his naked eye; an event which takes place only every few hundred years.

You are observers of variable stars, I am an observer of galaxies. Yet a very common ground between the two are the supernovae, and I thought that tonight I would tell you a little about galaxies, a little about supernovae. Supernovae are important to astronomers because they relate to a wide variety of phenomena which we want to study: the death of stars, the production and the recycling of heavy elements into the interstellar medium, the origin of cosmic rays, the origins of neutron stars; perhaps even black holes, the formation and evolution of supernovae remnants; perhaps even of quasars. They can be used to get distances of galaxies and to evaluate the Hubble constant; they may be important in stimulating star formation and may even affect solar system formation. Most directly, they are related to stellar evolution.

We live in a flattened galaxy of stars. All of the stars that you can see with the naked eye on a clear night are members of our galaxy, gravitationally bound to a distant center, and rotating about that center. It was just 50 years ago that astronomers learned what a galaxy is, and most of their understanding came from the observation that stars do not appear only singly. They appear in clusters, in clouds, and in groups. The first groups which were recognized formed patterns on the sky—constellations

(Fig. 1). Astronomers believe that the evolution of our galaxy is closely related to the evolution of individual stars. In the early expanding universe, there were atoms of hydrogen and probably helium, moving with random irregular motions. Somewhere an irregularity produced a re-

Figure 1. *The constellation of Cassiopeia, from the star atlas Uranometria of Bayer, 1603. The supernova of 1572, located in this constellation, was visible to the naked eye for two years. Bayer was so impressed that 31 years later he included the supernova in his atlas as the large, decorative star.*

gion of higher density, a protogalaxy, which gravitationally attracted more atoms. As the particles collided, the irregularities were damped out, the rotational motion became dominant, and the particles started collapsing to a plane. Stars which formed early in this history were formed at large distances from the plane, with elongated orbits reflecting the irregular motions of the gas from which they formed. These stars were composed mostly of hydrogen, with few metals. They are the halo stars and globular clusters, which we recognize as the oldest objects in our galaxy. As these stars evolve, they synthesize heavy elements in their interiors, and shed mass from their surfaces by stellar winds, or by shells from planetary nebula, and occasionally by supernovae explosions. New stars continue to form, with these younger ones formed closer to the principal plane of the galaxy, the plane where the gas is now located. These newer stars have a more complex composition, which includes heavier elements. Because repeated collisions had regularized the motions of the gas clouds, the newer stars have more circular orbits. In our galaxy star formation is going on now only in the plane, in the dusty gaseous regions which are so pretty to look at with the telescope.

Not all galaxies have evolved as ours did. Our nearest neighbors, the Magellanic clouds, satellites of our galaxy, are not strongly concentrated to a plane and do not have a massive central nucleus. They do, however, have an impressive population of hot, young stars. Star formation is going on at a very high rate in these galaxies. For some reason which we do not yet understand, these galaxies formed stars at a very slow rate in the past. They do have old red stars, but not many. Only now are they turning their large gas stores into stars. We do not know why star formation took place slowly in the past, or why it happens more rapidly at the present time in these systems.

I have spent this time describing galaxies because they are important to our understanding of supernovae. A supernova is a star which shows a sudden increase in brightness, becoming suddenly millions of times brighter than the Sun, and then decreases in brightness over the course of 50 days to several years. They can become brighter than an entire galaxy; then their brightness is over one billion times that of the Sun. Although supernovae are not all similar, they are all believed to be the end stage of a star that has converted to iron all of its chemical elements lighter than iron. Lacking other sources for producing energy and thus remaining stable, the star undergoes an energy crisis. The core collapses, and the outer envelope explodes into the surrounding interstellar medium. Spectra show that supernova envelopes move with velocities of 10,000 km/sec. Until recently, the complex broadbanded spectra were an enigma, but it is now suspected that many of the lines are due to the element iron. This is significant because of the role that iron plays in the star's nucleogenic evolution.

How many stars become supernovae? All? Few? In our galaxy, six or seven supernovae have been recorded in historical times. Images a thousand years old depicting supernovae sightings have been uncovered in the American southwest; manuscripts describing their occurrence exist in Chinese and Korean. A best guess gives five supernovae observed from Earth every 1000 years; the actual number of occurrences is at least 10 times higher.

Supernovae in external galaxies have been searched for fairly systematically since about 1936, when Fritz Zwicky of the California Institute of Technology and a group of about a dozen collaborators started using the 18-inch Schmidt telescope on Palomar Mountain to photograph many galaxies, each at two widely spaced times. They then compared the films to identify stars that were bright on one exposure and invisible on the other. From 1936 to 1973, 270 supernovae were found at Palomar; about 110 others were found by amateurs and professionals elsewhere. Zwicky calculated that the cost of the telescope construction, plus the films, plus the salaries, averaged out to less than $550 per supernova discovery. Surely a bargain, when we consider their value in teaching us about the evolutionary history of stars.

Supernovae leave gas remnants. They leave bubbles and lace-like filaments. They also leave a radio signature; a shell of compressed gas in the interstellar medium, which can be detected both by its hydrogen radiation and by its nonthermal synchrotron radiation arising from electrons that have been accelerated in the magnetic fields.

Supernovae also leave stellar remnants. Sometimes the remnant is a neutron star, as the residual supernova mass collapses and cools. The magnetic field also collapses; the stellar remnant may emit pulses of beamed electromagnetic radiation. We see it as a pulse when rotation carries it into our view. The most famous supernova remnant is the Crab Nebula, the residue of the supernova of 1054 AD. First seen as a remnant in 1731, it was not understood until 1940 as the remains of the supernova of 1054. The light from the remnant star at the center of the Crab Nebula pulses on and off 33 times per second. Like other pulsars, its rotation will slow down as it ages. At present, only two optical pulsars are known, the Crab and Vela, while 100 radio pulsars are known. The identification of the remainder is a problem.

How can you, observers of variable stars, find a supernova? Field glasses or a small telescope will do. Look at the intrinsically brightest Sc galaxies, Sc I galaxies. These are galaxies with small bulges, well-defined strong spiral arms, lots of high-mass young stars defining the arms. These are the best candidates. A high luminosity Sc galaxy has a supernova on average every 18 years; a high luminosity Sb has a supernova only every 50 years.

Perhaps Sc galaxies in the nearby Virgo cluster, in the Spring sky, would be a good place to start. Assume that a supernova stays bright for a month. If you look at a single galaxy once a month for 18 years, you might find "its" supernova. But if you look at 18 different Sc I galaxies once a month, you should find a supernova in one year! Happy hunting!

There is one other exciting facet of star formation that is just beginning to be understood by astronomers. In tenuous dust and gas clouds, like Orion, there are regions of much higher density in which hydrogen molecules are slammed onto each other by some giant shock. These higher density clumps gravitationally attract more particles, thereby encouraging the birth of a new star. Newly formed stars are embedded in a cocoon of opaque molecules—nature does not permit us to see the birth of new stars. By the time the gravitational pounding from the infalling matter has raised the temperature to a few hundred degrees, these regions are sources of intense infrared radiation; infrared observers are permitted to witness the birth of the next generation of stars. Now the question arises: where does the shock come from that creates the density high enough to start a new star? It may be from the other hot, young stars in the cluster, or it could be from the shock attending the supernova death of one of the older cluster stars. Thus, supernovae may do something even more spectacular than supply energy and heavy elements to the interstellar medium; they may initiate the formation of new stars. In fact, the high abundance of some elements offers evidence that a supernova exploded in the vicinity of the solar nebula as the sun was forming some 4 or 5 billion years ago. These heavy chemical elements, formed during the millions of years that the earlier star was evolving, became part of the atoms and molecules which were incorporated into the sun, the planets, the earth, and ourselves.

We still have much to learn about stars, galaxies, supernovae, and their interacting roles in the evolution of the universe. Your observations will help us to know more.

Dynamics of the Andromeda Nebula

1973

A ll the stars that can be seen with the unaided eye from the earth belong to our galaxy. They are members of a flat spiral system that rotates around a massive center some 10,000 parsecs from the sun. (One parsec is 3.26 light-years.) It is difficult to study the internal motions of the galaxy directly because we are located in its central plane, which is clogged with interstellar dust and gas that at visible wavelengths obscure the galactic center and most of the more distant stars. Thus, although we can study some of the motions of the galaxy from the observation of relatively nearby stars and from the radio waves emitted by distant clouds of hydrogen, in order to learn more about the dynamical behavior of galaxies we must turn to other systems. The nearest galaxy closely resembling our own is the Great Nebula in Andromeda, and its internal motions have recently been studied in considerable detail. Some of the results of these studies are quite unexpected.

On a clear night away from city lights in the Northern Hemisphere the Andromeda nebula (Fig. 1) is just barely visible to the unaided eye as a faint, elongated patch of light. It was described by the Persian astronomer Umar al-Sufi Abd-al-Rahman in the tenth century; it appeared on Dutch star charts in 1500. It was first observed with a telescope in 1612 by Simon Marius of Germany, who described it as resembling the light of a candle flame seen through translucent horn. In 1781 Charles Messier of France listed it as No. 31 in his catalogue of nebulous objects, and to this day the Andromeda nebula is also commonly known as Messier 31, or M31.

William Parsons of England, better known as the Earl of Rosse, began observing M31 in 1848 with his 72-inch speculum-metal reflecting telescope. His journal of observations was published some 40 years later, in

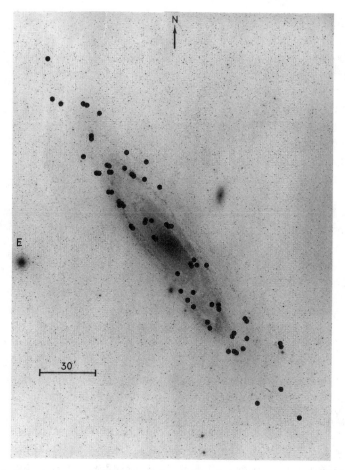

Figure 1. *The Andromeda nebulae (also known as M31), is the nearest spiral galaxy that resembles our own. We have obtained velocities for the regions indicated by the filled circles. Photograph from the Palomar 48-inch Schmidt telescope, courtesy of Dr. S. van den Bergh.*

1885. That same year a brilliant new star—a supernova—appeared near the center of M31. Ultimately this star served as a link in a chain of reasoning that established that spiral nebulas were not nearby clusters of stars or clouds of gas, but they were stellar systems outside our own. Another link was provided by Sir William Huggins, who obtained the first spectrogram of M31 in 1890, and by Julius Scheiner, who first discussed the spectrum of M31 in 1899. Scheiner recognized that the spectrum arose from the light of many stars rather than from a glowing cloud of gas.

Modern observations of M31 date from 1914, when V. M. Slipher, using the 24-inch refracting telescope at the Lowell Observatory in Flagstaff, Arizona, determined that the solar system and the center of M31 are approaching each other at a speed of 300 kilometers per second. It is now known that most of this observed velocity reflects the motion of the sun around the center of our galaxy. The sun is rotating around the galactic center at a velocity of about 250 kilometers per second in the direction of M31. If we could make observations from the galactic center, they would show that our galaxy and M31 are actually approaching each other at a rather modest speed; some 50 kilometers per second.

Soon after Slipher had made his observations, F. G. Pease obtained two spectra of M31 with the 60-inch reflector on Mount Wilson. For the first spectrum he lined up the slit of the spectrograph along the long axis of the tilted galaxy; for the second one he lined up the slit along the short axis. Each spectrum was the product of some 80 hours of exposure time spread out over a period of three months! The absorption lines of the stars in M31 were inclined at an angle in the spectrum taken along the long axis, but they were not inclined in the spectrum taken along the short axis. Pease correctly inferred that the shift of the lines was due to the fact that the galaxy was rotating.

Over the next three decades further observations of the rotation of M31 were conducted by Horace W. Babcock and Nicholas U. Mayall at the Lick Observatory, but the extreme faintness of the individual stars made the exposure times almost prohibitively long and made it impossible to determine velocities very far from the center of the galaxy, where the density of stars is low. In the 1950s and 1960s, however, the situation changed.

First, photography had advanced to the point where photographic plates sensitive in the red region of the spectrum had been developed. Walter Baade, working with the 100-inch reflector on Mount Wilson, detected faint nebulous patches on a red-sensitive plate during a long exposure he was making of M31 for another purpose. He inferred that these nebulous patches, now known as HII regions, were clouds of gas ionized by the ultraviolet radiation of hot stars. The red color of the patches arises from the fact that the gas and dust in M31 absorb the blue light of the galaxy more than red light. Moreover, a significant portion of the light from the HII regions is emitted at two lines in the red region of the spectrum: the line designated hydrogen alpha and a "forbidden" line of ionized nitrogen. (Forbidden lines arise when an electron within an atom drops from a state of higher energy that is metastable, or long-lived, to the ground state. In the diffuse gas of interstellar space an oxygen atom spends approximately 10^4 seconds, or some three hours, in such a metastable state before radiating a strong forbidden line in the visual region of the spectrum. Under normal conditions on the earth the atom would be deexcited by many collisions during that length of time and would never lose its energy by radiating a forbidden line.) During Baade's lifetime he identified 688 HII regions in M31.

We now know that young, hot Type O and Type B stars and their surrounding ionized HII regions are a major constituent of the spiral arms of galaxies. In the middle and late 1930s, however, Edwin P. Hubble of the Mount Wilson Observatory had been unable to detect any nebulous regions in M31 with photographic plates sensitive in the blue region. He had concluded that the absence of HII regions meant the absence of Type O and Type B stars in the galaxy. It was the availability of photographic plates sensitive in the red that made Baade's later observations successful.

A second change in the situation was that in the early 1960s electronic image-intensifiers, or image tubes, had come into their own. When photons of starlight enter an image-intensifier, they strike a photoemissive surface that ejects electrons. The electrons are accelerated and multiply, and finally they produce an amplified image of the star field or spectrum. Image-intensifiers in the early 1960s were comparable to photographic plates in their resolution and their ability to distinguish a signal from noise, and they could produce an image 10 times more quickly. Moreover, their relative gain was even greater in the red, where photographic plates are relatively insensitive and where the HII regions emit their principal radiation. Hence the study of HII regions in M31 by photographing the final image-intensifier phosphor was full of promise.

The spectrum of an HII region consists primarily of bright emission lines. Some are recombination lines of hydrogen and helium, which arise as an electron combines with an ionized atom and drops to lower energy levels. Also present in the spectrum are forbidden lines of un-ionized oxygen and lines of singly and doubly ionized oxygen (oxygen stripped of one or two electrons) and singly ionized nitrogen and sulfur. With a telescope of moderate size and an image-tube spectrograph (an image-intensifier attached to a spectrograph) it is possible to record the bright lines in the spectrum of an HII region in M31 in about an hour.

The velocity of each HII region along the line of sight results in a Doppler shift of the emission lines from their normal unshifted position. By measuring the location of the hydrogen-alpha line on the photographic plate to the nearest micron (thousandth of a millimeter) with respect to comparison lines put on the photographic plate at the telescope, one can obtain a velocity for each HII region. From the observed velocity one can compute the HII region's circular velocity around the center of M31. In principle one could observe the spectra of individual stars and obtain their velocities instead of the velocities of the HII regions. However, because the spectrum of a star is continuous, that is, spread out over all wavelengths, a star whose magnitude over all these wavelengths is equal to the magnitude of an HII region will require an exposure that is many times longer. HII regions emit almost all their light in just a few spectral lines. My colleague W. Kent Ford, Jr., and I exploited this fact when we chose to obtain spectra of these regions.

For the past six years, Ford and I have been obtaining spectra from selected areas within M31: from 70 individual HII regions that define the spiral arms, from the integrated starlight of the bulge of the galactic nucleus and from a disk of diffuse excited gas within the nuclear region. From these spectra we have mapped regions of differing velocity within M31 and have learned about the variation of the abundance of the chemical elements as a function of distance from the center of the galaxy. We have been working on the spectra of the HII regions since 1966, using a spectrograph that was designed and built in the Department of Terrestrial Magnetism of the Carnegie Institution of Washington.

The spectrograph has been attached to the 72-inch Perkins reflector of Ohio Wesleyan University and Ohio State University at the Lowell Observatory and to the 84-inch reflecting telescope at the Kitt Peak National Observatory near Tucson, Arizona. All the HII regions are too faint to be visible in telescopes of this size. In order to obtain a spectrum of each HII region we had to offset the telescope from brighter visible stars on the basis of positions that had been determined beforehand from long-exposure photographs.

We have not identified any emission regions closer to the galactic nucleus than 3,000 parsecs. On spectra taken of regions near the nucleus, however, a weak forbidden line of singly ionized nitrogen emitted by the diffuse gas does appear superimposed on the spectrum of the integrated starlight. For the gas we have been able to measure velocities to within a few parsecs of the galactic center. We plotted the velocities of the gas in the nucleus and of the HII regions against their distance from the center of M31. The most striking feature of the resulting rotation curve is a deep minimum in the gas velocities at about 2,000 parsecs from the center (Fig. 2.). If we assume that the motions we observe arise from particles of dust and gas moving in circular orbits in the gravitational field of M31, then it is possible to map the distribution of mass within the galaxy from the velocities. In this way we determined the distribution of mass within M31 with respect to distance from the nucleus. That curve too displayed a deep minimum at a radius of 2,000 parsecs from the galactic center. The minimum implies that there is a region of very low mass at that distance.

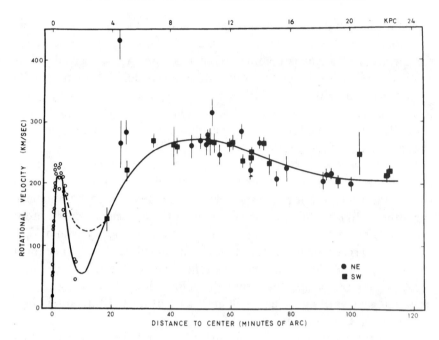

Figure 2. *Rotation velocities measured in M31, as a function of distance from the center. The solid line shows the adopted rotation curve.*

The total mass of M31, out to the last observed HII region some 24,000 parsecs from the galactic center, is 1.8×10^{11} times the mass of the sun. The form of the rotation curve near the nucleus does not affect the determination of the total mass or the distribution of the mass outside the nuclear region of M31. Even if the observed velocity minimum is due to some local disturbance, and does not indicate the overall normal circular gravitational velocity at that point, there is definitely a peculiarity in the velocities of the gas at a radius of 2,000 parsecs from the center.

If the only velocities observed are from stars or gas moving in normal circular orbits around the nucleus of the galaxy, then all the velocities we see along the short axis should be equal to the velocity of the center of M31; there would be no component of the rotational velocity along the line of sight. This is not, however, what the observations show. We have measured the velocities of the gas in the nucleus along the short axis. These measurements indicate that at some points very close to the nucleus, gas is flowing out of it at velocities that range as high as 135 kilometers per second. Only a very small amount of mass, however, is leaving the central region: less than 1% of the mass of the sun per year.

Ford and I were intrigued by the presence of this diffuse ionized gas near the nucleus of the galaxy, and so we extended our observations to nearby regions. The emission line of singly ionized nitrogen was difficult to detect above the background radiation from the stars. We attempted to increase our system's sensitivity by increasing the dispersion of the spectra from 135 angstroms per millimeter to 28 angstroms per millimeter. The result of this procedure is that the light from the background stars is spread out by a factor of five and is therefore dimmed, while the sharp forbidden emission line of singly ionized nitrogen is left unaltered. The technique was so successful that we were able to detect five emission lines from the gas where formerly we had seen only one. The only penalty for working at a higher dispersion is one of exposure time: even with our image-tube spectrograph, exposures as long as six hours were necessary.

From spectra taken at 16 different angles across the nucleus of the galaxy we have deduced how the diffuse nuclear gas is distributed. Within some 400 parsecs of the nucleus the gas is concentrated into a very flat disk, perhaps 25 parsecs thick. The disk is rotating with velocities that reach 200 kilometers per second about 200 parsecs from the nucleus. In addition to the rotation, gas is streaming outward from the nucleus, principally in two directions about 180 degrees apart near the short axis. The velocities of the streaming gas reach 135 kilometers per second near the far side of the short axis. Moreover, in locations where the spectrograph slit crosses dust patches in M31, the observed velocities suggest that clouds of gas and dust are falling into the central plane of the galaxy. This gas could have been shed by evolving stars, and it could contribute to the mass of the disk. Although this model simplifies some of the complex motions we observe, it does account for their major features.

In 1963 Bernard F. Burke, Kenneth C. Turner and Merle A. Tuve of the Department of Terrestrial Magnetism began studying the motions of the un-ionized hydrogen gas in M31 at the radio wavelength of 21 centimeters. More recent radio investigations have been conducted by Morton S. Roberts of the National Radio Astronomy Observatory and S. T. Gottesman, V. C. Reddish and R. D. Davies of the Nuffield Radio Astronomy Laboratories at Jodrell Bank in England. Since the angular resolution of a radio telescope is lower than the resolution of an optical telescope, the actual detail that can be resolved in M31 is much less. In studies covering all of M31 to date, the diameter of the radio telescope beam on the sky has been 10 minutes of arc or greater. That corresponds to an ellipse some 2,000 parsecs wide by 9,000 parsecs long on M31 because of the fore-shortening produced by the fact that we see the galaxy tipped only 13 degrees from edge-on. Our spectrograph slit covers about five parsecs by 400 parsecs on the galaxy. In spite of the difference in resolution, however, the agreement between the rotation curves resulting from the radio studies and those resulting from the visual studies is impressive for distances greater than 3,000 parsecs from the center of M31. (Closer than 3,000 parsecs from the nucleus there is too little un-ionized hydrogen to be detected easily by radio telescopes.)

The radio-wavelength observations have extended our knowledge to the outer limits of M31. I have mentioned that at visible wavelengths we have been able to determine velocities within the galaxy only out to the outermost known HII region 24,000 parsecs from the nucleus. In 1972, Roberts and Robert Whitehurst of the University of Alabama extended the rotation curve out to 34,000 parsecs from the center with 21-centimeter observations. They found that the rotational velocity of un-ionized hydrogen remains constant at 200 kilometers per second between 24,000 and 34,000 parsecs from the nucleus. The mass contained between those limits is equal to 10^{11} times the mass of the sun, yielding a total mass for M31 out to a radius of 34,000 parsecs of 3×10^{11} times the mass of the sun. Since we still know very little about the boundaries of galaxies in general, this extension of the diameter of M31 is a matter of some importance.

So far I have discussed only the motions of the gas within M31 and by inference the motions of the hot young stars that ionize the HII regions. The study of the motions of stars themselves is more difficult. Part of the reason is, as I have mentioned, that the individual stars are faint and that their spectra are continuous. We do see broad, diffuse stellar absorption lines in the continuous spectra from regions near the nucleus. Indeed, it is possible to measure the velocities of stars from these lines, although the measurements are less precise than those that can be made with the sharp emission lines from the interstellar gas. Moreover, the analysis is complicated by the fact that stars near the nucleus of the galaxy are distributed not in a flat disk as the gas is but in a nearly spherical bulge. Hence we must obtain the spectra across a long projected path. At large distances from the galactic nucleus the velocities of individual stars are still too difficult to obtain.

We have measured the velocities of the stars on both the long and the short axes near the nucleus of M31 with a single absorption line. We anticipated that the velocities would indicate that the stars are following simple circular orbits around the center of the galaxy, and that they would exhibit none of the complexities of the gas velocities. This, we have found, is not the case. The general features of the stellar motions resemble those of the gas motions. There is a steep gradient in the velocities of the stars across the galactic nucleus reaching a minimum at 2,000 parsecs, and the stars seem to be moving outward along the short axis. In 1939 Horace Babcock had observed a minimum in the stellar velocities near 2,000 parsecs, but his observations had never been confirmed or widely accepted.

For several reasons these results are both unexpected and not understood. On the basis of our present understanding of stellar evolution it is believed the stars whose spectra we are observing near the nucleus of M31 are some four billion years old. In contrast, the irregularities in the motions of the gas should smooth out in less than 10 million years. Thus it is not clear how the old stars, which during their long life should have moved several times around the galaxy, can have the same motions as the young gas. Moreover, although it is relatively easy to work out mechanisms that impart irregularities to the gas motions, it is difficult to do the same for stars. Our present suggestion is that we are observing stars traveling in noncircular orbits that are dynamically stable. One possibility is that some of the mass of the nucleus of M31 is distributed asymmetrically. The inner spiral arms of the galaxy are decidedly asymmetrical, but it is unlikely that their mass is great enough to distort the orbits of the stars. Perhaps there are resonances, or other cooperative effects, that stabilize the orbits (as gravitational theories of galactic structure now predict). Stars are known to shed mass as they evolve; it is possible that the gas we observe has come from old stars. Detailed understanding must await further observations.

Beyond the region in M31 where the velocities of stars and gas fall to a deep minimum at 2,000 parsecs from the galactic nucleus, the HII regions have circular velocities that rise to 250 kilometers per second at 10,000 parsecs from the nucleus and remain at that velocity out to 24,000 parsecs. We can deduce a remarkably similar pattern of velocities for our own galaxy. A flattened disk of un-ionized hydrogen extends several hundred parsecs from the center, rotating with velocities of up to 200 kilometers per second and expanding at velocities as high as 135 kilometers per second. At a distance of 800 parsecs from the galactic nucleus both the velocity and the density of the gas are very low. Near the position of the sun, 10,000 parsecs from the nucleus, the stars and the gas are rotating around the nucleus at a velocity of some 250 kilometers per second. That velocity decreases to approximately 175 kilometers per second at a distance of 24,000 parsecs. The overall resemblance between these figures and those for M31 implies that the distribution of mass in our galaxy is similar to that in M31.

Spectrographic studies of M31 that we have carried out with C. Krishna Kumar of the Department of Terrestrial Magnetism also provide information about the relative abundances of the chemical elements in the galaxy.

The abundances can be determined from the strength of the spectral lines emitted by atoms of the elements with respect to the strength of the hydrogen-alpha line. The abundances, like the velocities, vary with distance from the nucleus. In the diffuse gas of the nucleus the spectral lines of oxygen, nitrogen and sulfur are all stronger than the hydrogen-alpha line. Some of the HII regions 4,000 parsecs away from the nucleus also show these anomalously strong lines, although in the normal HII regions at that distance the lines of all three elements are weaker than the hydrogen-alpha line.

Between 5,000 and 15,000 parsecs from the nucleus the nitrogen lines decrease in strength with increasing radius and the oxygen lines increase in strength. We infer from this observation that the abundance of nitrogen is decreasing with respect to the abundance of hydrogen by a factor of about one-half with increasing distance from the nucleus. The abundance of oxygen is also decreasing, but by a small factor: perhaps 1/1.2. The decrease in oxygen has a curious result, because under the conditions of temperature and density in an HII region, radiation by oxygen atoms acts to cool the gas in which they are embedded. Therefore, a decrease in oxygen leads to an increase in temperature and to a strengthening of the line of doubly ionized oxygen with respect to the line of hydrogen alpha.

In our galaxy we can observe HII regions only near the sun, that is, at a distance of 10,000 parsecs from the galactic nucleus. In these regions the strongest spectral line is always hydrogen alpha. Leonard Searle of the Hale Observatories has studied giant HII regions in external galaxies other than the Andromeda nebula, and he finds an even larger variation in the strength of the lines of oxygen and nitrogen than Ford and I see in M31. A decrease in the abundance of oxygen, nitrogen and sulfur with increasing distance from the galactic nucleus is probably a general feature of normal spiral galaxies. Perhaps the interstellar gas near the nucleus has been enriched over several generations in the formation of stars. During its lifetime a star transforms by thermonuclear reactions some of the hydrogen in its interior into heavier elements. These elements are returned to the interstellar gas if the star explodes at the end of its life cycle, resulting in an interstellar gas that is richer in heavy elements than it was initially.

Our study of M31 has been a satisfying one. There have been surprises, such as the existence of the disk of gas in the galactic nucleus and the complexity of its motions. There have been puzzles, such as the motions of the stars close to the nucleus. There has been controversy, such as the interpretation of the minimum in the velocities observed near the radius of 2,000 parsecs. Overall, however, we have been able to determine the details of motions in a galaxy much like our own. These results suggest new problems to examine in order that we may better understand the dynamics of spiral galaxies in general and of our own galaxy in particular.

The Peculiar Galaxy NGC 1275[a]

1976

A fundamental question of cosmology concerns the arrangement of matter in the universe. Is most of the mass of the universe contained in galaxies, or is there a sizable fraction of mass outside the galaxies? A second question concerns the significance of the division of galaxies into two major types. Why do some galaxies have flattened disks, contain gas, dust, and young stars in significant quantities, and usually exhibit spiral structure; while other galaxies are spheroidal, dust- and gas-free, generally structureless, and contain principally older, evolved stars? Answers to these questions can be approached by (1) statistical examination of large galaxy samples, and (2) studies of masses of fairly nearby typical galaxies (deduced from a detailed study of their velocity fields). It is the latter approach which has occupied part of our research efforts over the past few years.

During the last decade, significant progress has been made in observing the velocity fields and hence the masses of spiral galaxies, as measured from the emission lines in their spectra. These emission lines arise in the excited gas in each galaxy. Although the gas constitutes in general only a small percentage of the mass of the galaxy, it is assumed that the gas is in dynamical equilibrium in the gravitational field of the stellar population, and can thus be used as a tracer of the underlying velocity pattern.

For gas-free stellar populations such as are found in the bars of barred spiral galaxies or in elliptical or S0 galaxies, velocity data are almost non-existent and mass data are meager. During the past year, we have directed our observations toward these difficult objects, and have obtained spectra of sufficiently high quality so that we could measure the velocities of the stellar populations from the absorption lines in the integrated galaxy spec-

---

[a] With W. K. Ford, Jr. and C. J. Peterson.

tra. We report here on NGC 1275, an exceedingly peculiar galaxy whose nature has long been a puzzle (see Fig. 1).

NGC 1275 (Perseus A) is one of the most peculiar objects in our region of the universe. Twenty years ago, Minkowski discovered in its spectrum two sets of emission lines, one set with a redshift $V \sim 5200$ km s^{-1} and a second with $V \sim 8200$ km s^{-1}. Since that time NGC 1275 has been interpreted as either an exploding galaxy or a collision between two galaxies. In order to study the underlying stellar population and the excited gas in NGC 1275, an extensive body of spectroscopic and photographic material has been obtained at the 2.1-m and 4-m telescopes of Kitt Peak National Observatory. Analysis of this material in collaboration with Professor J. H. Oort leads to the following conclusions:

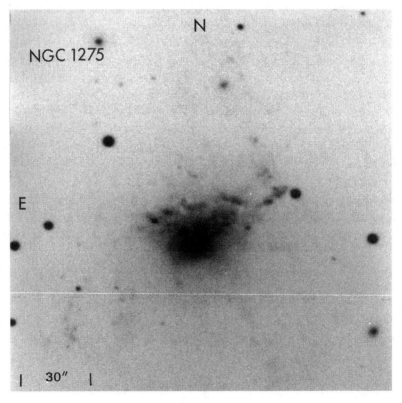

Figure 1. NGC 1275, from a Kitt Peak National Observatory 4-m telescope plate taken by Dr. S. Strom. The knots at the north of the galaxy all have velocities about 3000 km/s higher than the underlying spheroidal galaxy.

1. The luminosity profile of the stellar continuum shows that the main galaxy is spheroidal. A strong absorption line spectrum is seen at $V = 5200$ km s^{-1}, which agrees with the velocity of the extensive filamentary material surrounding the galaxy. South of the nucleus is a strong A-type absorption line spectrum; to the north the Balmer lines are obscured. Emission lines of H, O, and N are also seen.

2. Emission patches at $V = 8200$ km s^{-1} over the northern part of NGC 1275 are associated with obscuration that hides the stellar continuum of the underlying galaxy. The pattern of velocities in the high-velocity system shows a fairly smooth transition from velocities near 8100 km s^{-1} in the east to velocities of 8400 km s^{-1} in the west. This pattern is consistent with that expected from a late-type galaxy rotating with a maximum velocity of $V_{max} = 150$ km s^{-1}/sin(inclination angle). In addition, the emission line ratios seen in this high velocity gas are typical of large, low-density HII regions observed at large distances from the nucleus in late-type galaxies. No stellar absorption lines have been found associated with this system. This high-velocity gas is in front of the low-velocity system.

NGC 1275 is at the center of the Perseus cluster of galaxies. A count of galaxies in the central region of the cluster suggests that the probability of finding a spiral galaxy within 30 arcseconds of the nucleus of NGC 1275 has the reasonably high value of 1/30.

3. NGC 1275 is a strong radio source as well as a strong infrared and strong x-ray source. This, in addition to the optical appearance of the low-velocity filamentary material, reminiscent of the Crab Nebula supernova remnant in our own galaxy, implies that the object has undergone a violent event in the past. The A-type stellar spectrum indicates that a burst of stellar formation was associated with this event. It is likely that the time scale required for the dispersal of material into the low velocity filament system is longer than the time scale over which the two galaxies may have been in gravitational interaction (the two objects are still approaching each other); thus, the presence of the second galaxy cannot have triggered the explosion in the underlying galaxy. Hence our interpretation of the NGC 1275 phenomenon requires both an explosive event in the elliptical galaxy and a high-excitation ragged irregular (late-type) intervening galaxy. While it is unlikely that the two main galaxy masses are interacting, it is still possible that an energetic nuclear source in the underlying elliptical may be the source of excitation for the high-excitation spectrum observed in the late-type galaxy.

Moderately peculiar galaxies are commonplace in the universe. Severely peculiar galaxies are less so, but the attempt to understand them has led some astronomers to question whether yet unknown laws of physics govern their behavior. Our observations of NGC 1275 show that individual properties, i.e., the luminosity profile of the underlying galaxy, the observation of the northern part, the velocity and line-intensity pattern of the high velocity system, all can be understood in terms of previously known characteristics of galaxies. Such is the progress of astronomy. Only when detailed studies reveal characteristics that are outside the realm of previously observed properties of galaxies can we properly appeal to an unknown physics.

UGC 2885, *the Largest Known Spiral Galaxy*

1980

H ow large a galaxy appears in a telescope or on a photographic plate, (what we call its angular size) depends on two factors: its distance and its intrinsic linear size. Until recently, astronomers generally did not describe galaxies in terms of their linear dimensions, due to uncertainties in determining their distances. But recent studies have shown that galaxies come in an enormous range of diameters, and to ignore this range is to neglect one of the few galaxy parameters which can be evaluated. For UGC 2885 (galaxy number 2885 in the Uppsala General Catalogue of 12,921 galaxies in the northern equatorial hemisphere, compiled by Peter Nilson) its diameter is its claim to fame: it is 250,000 parsecs (almost 1 million light years) wide, making it the spiral galaxy with the largest known diameter (Fig. 1).

On the sky, UGC 2885 appears to be about 5.5 minutes of arc across, or as large as the spiral galaxies nearest to our own. However, its redshift is five times as large as the redshifts for spirals in Virgo, making it five times as far away, and five times as large. Both by its position on the sky and its redshift, it is identified as a suburban member of the extended Perseus Supercluster of galaxies, with a velocity of recession near 6000 km per sec. Because we view UGC 2885 in a direction not far from the plane of our Galaxy, it is obscured by the interstellar gas and dust within our Galaxy. Hence it is possible that its true dimensions are even larger than those deduced.

UGC 2885 is an attractive two-armed spiral with a small bright nucleus which appears almost stellar, and with bright emission knots observed even near the extremities of its arms. The presence of these ionized gas clouds surrounding hot, young stars implies that normal star formation is going

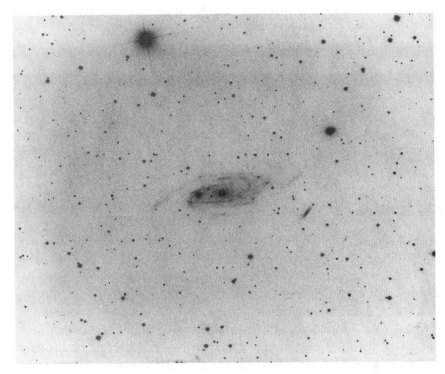

Figure 1. *A negative print of the galaxy UGC 2885 reproduced from a plate taken by Bruce Carney at the Kitt Peak 4-meter telescope. It is a 30-minute exposure on a IIa-O plate through a GG385 filter.*

on even at these large distances from its nucleus. Moreover, the chemical abundances in the distant regions do not appear remarkably unusual. Active star formation at such extreme distances from the nucleus is surprising in comparison with our own galaxy, where astronomers have traditionally concluded that star formation ceases at distances beyond the orbit of our Sun, about 10,000 parsecs.

It is its internal dynamics, coupled with its enormous size, that makes UGC 2885 remarkable. As in all disk galaxies, the stars and gas in UGC 2885 are orbiting the nucleus in circular paths. For UGC 2885, circular velocities are near 250 km per second just beyond the nucleus, and increase to near 280 km per second at the limits of the optical image.[a] Rota-

---

[a] Here, "limits" of the image means that the surface brightness has fallen to 25th magnitude per square arc second, a convenient measure for discussing the outer brightness of galaxies. All dimensions are calculated using a Hubble constant of 50 km per second per Megaparsec.

tional velocities remain high in all disk galaxies because of the presence of significant mass at large distances from the nucleus. This means that the mass is not centrally condensed in a disk galaxy, as it is, for example, in the solar system. In the solar system, rotational velocities decrease with increasing distance from the Sun: Mercury, at 0.39AU, orbits the Sun with a velocity of 47.9 km per second, while Pluto, 100 times farther at 39.5AU, orbits with a velocity one-tenth as large, 4.74 km per second. This Keplerian decrease in velocity (so called because its basic mathematical form was first described by Kepler) arises because essentially all of the mass in the solar system is located in the Sun. In contrast, the lack of any decrease in stellar velocities for stars at large distances from the nucleus in a disk galaxy tells us that the mass density is falling very slowly; so slowly that the total mass within the galaxy is not converging to a limiting mass at the limit of the optical galaxy. Still more mass, in the form of dark matter, must exist beyond the luminous galaxy. By dark matter we mean matter whose luminosity, per unit mass, is much lower than the normal stellar material. For extragalactic observers, determining the distribution (halo? disk?) and the physics (mini-black holes? Jupiters? bricks?) and studying the dynamics of this matter is an important future goal.

For UGC 2885, the rotation period is 2 billion years at a distance of 125,000 parsecs from the nucleus. The outer regions have therefore undergone fewer than 10 revolutions since the origin of the Universe. (This number is independent of the choice for the Hubble constant, for both the age of the Universe and the linear dimensions of the galaxy scale inversely with the Hubble constant.) Yet with even so few rotations, the arms are smooth and well developed, and there are no large-scale irregularities in the velocity field (or flow pattern) of the galaxy. Large-scale velocity regularity coupled with few revolutions means that a well-ordered global spiral pattern must be established soon after galaxy formation. It cannot be the product of smoothing introduced by many differential rotations. Hence large disk galaxies with global spiral patterns, such as UGC 2885, put important constraints on models of galaxy formation and evolution.

The mass within the optical image of UGC 2885 is about 2 trillion times the mass of our Sun. This is as great a mass as is known for any spiral galaxy. The total mass, the sum of the luminous matter and the nonluminous matter beyond the optical image, is unknown. But judging from the surprises which astronomers have uncovered in their studies of spiral galaxies, much more remains to be learned before we can state with certainty how the total mass is distributed in this or any other galaxy.

NGC 3067[a]

1981

T here is now compelling evidence that rotational velocities in spiral galaxies remain constant or increase slowly with increasing nuclear distance, out to the limits of the optical image. These observations imply that significant mass of low luminosity exists at large nuclear distances in spiral galaxies. They also suggest that knowledge of the character of outer disks of galaxies will be fundamental in understanding the distribution of mass in the universe.

It is not easy to devise observations that will determine the extent of such low-luminosity matter and the corresponding rotational velocities. One possibly successful approach is to use a background quasar as a continuum source against which interstellar absorption from the nonluminous portion of the intervening galaxy can be detected. The galaxy-quasar pair NGC 3067-3C 232 is one of the few known pairs whose geometry makes it useful for such a study.

The observational history of the NGC 3067-3C 232 pair can be summarized as follows. The quasar 3C 232 (with redshift $z = 0.513$) is a background object which appears on the sky several optical radii from NGC 3067, the optical galaxy. The galaxy is classified Sb III in the Revised Shapley–Ames Catalog [1]. A plate of the galaxy-quasar field taken at the Lowell Observatory 42-inch telescope is reproduced in Fig. 1. Haschick and Burke discovered a very narrow (< 5.5 km/sec) absorption line of neutral hydrogen at $V = 1418 \pm 2$ km/sec in the spectrum of the quasar, as well as broad hydrogen emission from the galaxy extending from $V = 1310$ to $V = 1640$ km/sec. [2]. From high-dispersion optical spectra of the quasar, Boksenberg and Sargent discovered the H and K lines of Ca II arising from the foreground galaxy [3].

---

[a] With W. Kent Ford, Jr.

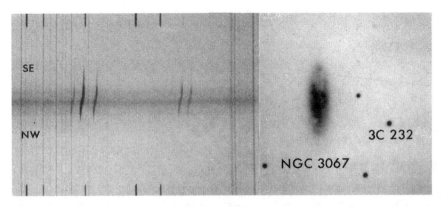

Figure 1. *(left) A spectrum of NGC 3067 taken at the Kitt Peak 4-meter telescope. The strongest galaxy lines are due to hydrogen, nitrogen, and sulfur. Orbits of stars and gas in the galaxy shift the emission toward the red spectral region (to the right) on the SE, and toward the blue (to the left) on the NW, with respect to the central velocity of NGC 3067. The weak straight lines crossing the spectrum are emission lines from the Earth's atmosphere. (right) An image of NGC 3067 taken with the Lowell 42-inch telescope plus Carnegie image tube. Neutral hydrogen gas extremely far out in NGC 3067, invisible optically, is observed in front of the very distant quasar, 3C 232.*

It is therefore well established that the line of sight to 3C 232 passes through gas associated with the foreground galaxy NGC 3067, gas well beyond the detectable optical image. The question posed by these observations is where in the galaxy the absorption originates. If from a coronal cloud, the cloud is more than 14 kpc above the plane. This distance is greater than the optical radius of the galaxy, 9.6 kpc ($H = 50$ km sec^{-1} Mpc^{-1}). But if the cloud originates in the disk, and is moving in a circular orbit viewed at an inclination of 68 degrees (the inclination of the optical galaxy), then the cloud tells us that the disk extends at least to 39 kpc, which is four times the optical radius.

We have determined a rotation curve for NGC 3067, to see what constraints the dynamical information can place on the interpretation of the absorption measures. Velocities come from two spectra of NGC 3067 (Fig. 1) taken with the Kitt Peak 4-meter RC spectrograph, a spectrograph which incorporates a Carnegie image tube. The rotational velocities shown in Fig. 2 come from Hα and [N II] λ6583 lines on each plate, reflected about the central velocity. The rotation curve is smooth and gently rising; from 3 to 7 kpc, the velocities increase linearly at the rate of 4.5 km sec^{-1} kpc^{-1} to a maximum velocity of 148 km sec^{-1}.

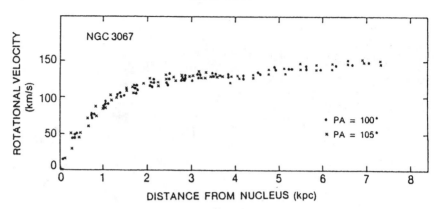

Figure 2. *Rotational velocities in the plane of the galaxy NGC 3067 as a function of distance from the nucleus. Note the slight increase in velocity with increasing distance beyond 3 kpc.*

If we assume that the absorption observed at $V = 1418 \pm 2$ km/sec along the line of sight to the quasar arises from a neutral hydrogen cloud orbiting with circular velocity in the plane of NGC 3067, then we can determine the radial distance and the orbital velocity of the cloud. Using the conventional geometrical expressions for deriving the circular velocity from an off-axis velocity, and the parameters adopted for NGC 3067, the radial distance of the absorbing gas is calculated to be 39(–4, +8) kpc; the rotational velocity is 335(–100, +125) km/sec. Thus, if the absorption arises from disk gas, the rotation curve must continue to rise from its final measured velocity of 148 km/sec at $R = 7.3$ kpc to $V = 335$ km/sec at $R = 39$ kpc. It is striking that this velocity is in accord with the velocity predicted ($V\sim 290$ km/sec) if the velocity gradient observed from 3 to 7 kpc continues linearly to 39 kpc. The listed uncertainties come from a consideration of the range of velocities and viewing geometry permitted by the observations. This V, R pair (335 km/sec, 39 kpc) is also consistent with values of V, R that we have observed for a sample of Sb galaxies with a wide range of luminosities.

In summary, these optical observations have added no new evidence as to the location of the cloud seen in absorption around NGC 3067. But they do force an interesting constraint. If the absorbing cloud is disk gas, then the rotational velocities in NGC 3067 continue to increase at least to 39 kpc, in accord with the velocity gradient seen in the optical disk, and in accord with properties of more luminous, more massive Sb galaxies. This observation has important implications for the distribution of mass in disk galaxies and in the universe.

The only way that circular velocities at $R = 39$ kpc can be higher than those at 9 kpc is for most of the mass of the galaxy to be located beyond 9 kpc, i.e., beyond the optical galaxy! The cloud at 39 kpc feels an interior dynamical mass of 1.0×10^{12} solar masses; only 5% of this is contained within the optical image, $R < 9.6$ kpc. Thus, 95% of the mass, but very little of the luminosity, comes from the region beyond the optical disk but within the radial distance of the absorbing cloud. These values require that the ratio of dynamical mass to optical luminosity, M/L, have a value greater than 100 at large radial distances; normal stellar matter has M/L ratios less than 10. Thus, NGC 3067 has 20 times more mass than would be deduced from a study of the optical galaxy alone. Once again we conclude that luminosity is a poor indicator of mass. The existence of heavy halos—to stabilize disk galaxies and to produce the flat or rising rotation curves that are observed—has been predicted earlier. But this study of NGC 3067 gives observational evidence (other than by inference from the gravitational effects) that these heavy halos do exist.

Do other galaxies have massive invisible halos? Only very special circumstances like this quasar-galaxy pairing may make it possible to answer this question by direct observation. We will continue to search out other candidates for similar study. Ultimately, such studies may tell us if we live in a low-density universe that will continue to expand forever, or if we live in a high-density universe, with much of the matter detectable at present only by its mutual gravitation. In such a universe the expansion may ultimately slow down, halt, and reverse if the density is sufficiently high. We are presently taking only the preliminary steps toward observations that can answer these questions.

References

1. A. Sandage and G. A. Tammann, *A Revised Shapley–Ames Catalog of Bright Galaxies* (Carnegie Institution of Washington Publ. 635, 1981).
2. A. D. Haschick and B. F. Burke, Neutral hydrogen absorption in the spectrum of the quasar-galaxy pair 4C 32.33/NGC 3067, Astrophys. J. (Lett.), **200**, L137–L140 (1975).
3. A. Boksenberg and W. L. W. Sargent, The existence of Ca II absorption lines in the spectrum of the quasar 3C 232 due to the galaxy NGC 3067, Astrophys. J., **220**, 42–46 (1978).

S0 Galaxies with Polar Rings[a]

1982, 1983

I n medicine, the study of rare pathological cases often leads to an improved understanding of whole classes of more normal specimens. Similarly, our chance discovery of a highly peculiar galaxy resembling a spindle surrounded by a detached ring (Fig. 1a) has not only revealed that such objects form a new small class, but has also led to significant new insights about the much larger class of "spindle" galaxies and even about disk galaxies in general.

Last year we reported the discovery of the faint anonymous galaxy A0136-0801, a member of a rare, new class of galaxies that appear to have spindle-shaped bodies girdled by luminous rings. Our observations showed that the apparent spindle of A0136-0801 is, in fact, a *disk* of old stars seen nearly edge-on, and that the luminous ring consists of young stars, gas, and dust orbiting at nearly right angles to that disk, i.e., over its poles. This unusual orientation of the ring allowed us to probe, for the first time, the vertical gravitational potential of a disk galaxy out to great heights. We found that the dark halo of A0136-0801 extends well beyond the visible edge of the embedded disk and is more nearly spherical than flat. Thus, through a quirk of nature that created this strange system, we gained valuable new insights into the structure of dark halos associated with disk galaxies.

The origin and evolution of S0 galaxies with polar rings are as interesting as their structure. The existence of two nearly orthogonal orbital planes can hardly be due to a single formation process; rather, it suggests a second event. But what was this second event, and what has been its subsequent evolution?

First, we note that ringed S0 galaxies are rare. From the eight or nine examples now known and an unsuccessful search of ten fields of the SRC

---

[a] With François Schweizer (first author) and Bradley C. Whitmore.

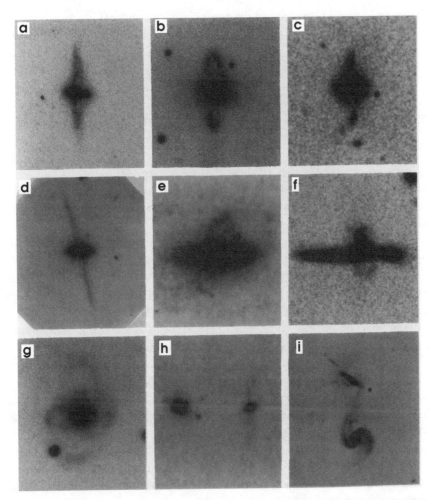

Figure 1. *Eight galaxies with polar rings and one related object: (**a**) A0136-0801, a spindle galaxy girdled by a polar ring of gas, dust, and young stars. The spindle galaxy is a rotating disk of stars seen almost edge on. The gas, dust, and stars in the ring are also rotating but in a plane almost perpendicular to the disk. Image from a 4-meter Cerro Tololo Inter-American Observatory plate taken by Dr. F. Schweizer. Galaxies **a**, **b**, and **c** illustrate a sequence of stellar disks seen increasingly face on. Panel **i** illustrates the mass transfer that often takes place when two galaxies interact tidally. Material from the spiral disk (bottom) is wrapping itself around the companion (top) and may eventually form a polar ring. [Photographs **d** and **i** from H. Arp; **c**, **e**, **f**, and **h** are reproduced from the Palomar and SRC(J) Sky Atlases.]*

Southern Sky Survey for further specimens, we estimate that only about three out of 10,000 field SO galaxies possess such rings. A likely explanation for these rare rings is the occasional infall of a small gas-rich galaxy in a nearly polar orbit around the pre-existing larger disk galaxy. For *any* infall angle, the well-known process of dynamical friction will inevitably lead to a slow spiraling in and disintegration of the smaller body due to frictional forces experienced in the halo. But for most infall angles, precession and subsequent collisions of the strewn-out gas clouds will force them to settle into the plane of the main disk within a time much shorter than the age of the universe. Only material in orbits nearly perpendicular to the main disk will precess so slowly as to survive for a time span comparable to or longer than the age of the universe. Calculations by G. F. Simonson of Yale University show that such long-lasting orbits will exist within typically about $10°$ from the poles. It seems more than just a coincidence, then, that all ringed SO galaxies found to date possess rings within roughly that angle from the poles.

The desire to understand the formation and evolution of these intriguing objects led us to undertake a survey of polar-ring galaxies [1]. We hoped that such a survey might allow us to address the questions: (1) What fraction of all galaxies possess polar rings? (2) Do the rings form around certain types of galaxies preferentially? (3) Does the presence of a polar ring depend on the environment of the galaxy? (4) Do some objects contain clues about the formation and evolution of polar rings?

Ideally, such a survey ought to be made by visually searching the approximately 1500 photographs of the combined northern and southern Sky Atlases. This task would consume several years, since galaxies with polar rings are rare and hence, on average, distant and faint. Fortunately, the Sky Atlases have been thoroughly searched for peculiar galaxies in recent years, and we could reduce our task to perusing available catalogs. We learned to concentrate on certain types of peculiarities, and inspected on the Sky Atlas photographs of each object having a promising description. After sifting through several hundred candidate objects, we ended up with a sample of 22 disk galaxies with polar rings.

A detailed statistical analysis of this sample has yielded two main results. (1) Polar rings occur in 0.2% of all galaxies outside of rich clusters; this fraction is an order of magnitude larger than we estimated originally, but still implies that occurrence is rare. (2) Polar rings occur preferentially in galaxies of Hubble type SO, which are disks of old stars nearly devoid of gas and dust. This type-preference seems highly significant, since only one-tenth of all field galaxies are type SO, whereas about two-thirds of the known polar-ring galaxies belong to this type. The formation process of polar rings that we propose below accounts quite naturally for the observed type-preference.

Besides providing vital statistics, our sample of polar-ring galaxies also solved the earlier puzzle of why the first eight specimens all appeared spindle-shaped. On the one hand, comparisons of the rotation and random motions measured in three of them [2,3] pointed unequivocally to their being S0 disks viewed nearly edge-on. Yet it seemed puzzling that all eight should share this same orientation. The enlarged sample now contains several galaxies that are clearly seen more face-on, just as one would expect if they all are disks. The top row of Fig. 1 illustrates a sequence of polar-ring galaxies with stellar disks seen increasingly face-on.

The apparent preponderance of systems with edge-on orientations is caused by two selection effects. First, a polar ring is more easily detected protruding from an edge-on S0 disk than projected against a face-on disk (compare Fig. 1a with Fig. 1c). Second, the increased surface brightness of an edge-on S0 disk over a similar face-on disk makes the former visible out to greater distances. Thus the results of our survey support the conclusion that the spindles are only apparent. Nature clearly prefers to form oblate galaxies; prolate galaxies, if they exist, have yet to be found.

A discovery from our survey is that the majority of polar-ring galaxies are components of binary galaxies or small groups. This is true, for example, for five of the eight specimens shown in Fig. 1 (b, c, f, g, and h). The companions invariably are disk galaxies themselves. Hence, the presence of polar rings depends on the environment in the sense that rings seem to be favored by the presence of neighboring disk galaxies.

The explanation is, we believe, that disk galaxies can exchange mass during close encounters. Simulations of such encounters in binary galaxies show that mass transfer induced by tidal forces is a common occurrence [4]. Occasionally, nature lets us witness the process directly: Fig. 1i shows NGC 3808, where material is being torn from the main spiral by the companion (to the top) and may soon be forming a polar ring.

As emphasized above, the presence of rings rotating at nearly right angles to the underlying galaxies implies that they formed after the main galaxy, i.e., in a "second event." To explain their formation in *isolated* galaxies, we had to postulate the disintegration and accretion of a satellite galaxy formerly in a near-polar orbit. The discovery that a majority of polar-ring galaxies occur in pairs and groups suggests that the second event is often less dramatic than a cannibalistic ritual: robbing a neighbor may do.

Why are so many polar-ring galaxies of type S0? The mentioned accretion mechanisms provide a natural explanation. Inevitably, accretion occurs in all types of galaxies and at random angles. However, only gas accreted within about 20° from the poles can form a semi-stable ring that survives for many billions of years. The ultimate fate of a forming ring must then depend on the relative amounts of gas present in the ring and in the main plane of the galaxy. If the latter gas dominates, it will sweep up the polar gas and force it to settle into the main plane; this is likely to occur in the gas-rich Sa–Sc spirals, whence they lack polar rings. If, on the other hand, the polar gas dominates, then it can complete forming a ring; this is likely to be the case in the gas-free S0 galaxies. Therefore, the preponderance of type S0 among polar-ring galaxies is a simple consequence of the lack of resistance encountered by the polar gas in the S0 disk.

One fascinating aspect of polar rings is that they are signatures of the past. Just as paleontologists are reconstructing the evolution of species from fossils, astronomers have now begun unraveling the evolution of galaxies from surviving signatures. The polar-ring galaxies confirm the picture that galaxies do not evolve in splendid isolation but rather interact, feed on each other, and have reached their present-day shapes through a long chain of evolutionary steps.

References

1. F. Schweizer, B. C. Whitmore, and V. C. Rubin, Colliding and merging galaxies. II. S0 galaxies with polar rings, Astron. J. **88**, 909–925 (1983).
2. P. L. Schechter and J. E. Gunn, NGC 2685: spindle or pancake?, Astron. J. **83**, 1360–1362 (1978).
3. P. L. Schechter, M.-H. Ulrich, and A. Boksenberg, NGC 4650A: the rotation of the diffuse stellar component, Astrophys. J. **277**, 526–531 (1984).
4. A. Toomre and J. Toomre, Galactic bridges and tails, Astrophys. J. **178**, 623–666 (1972).

A Pair of Noninteracting Spiral Galaxies[a]

1983

D uring the past few years, H. Arp has identified galaxy pairs that are seen close together on the sky, and has suggested that these pairs are physically connected and interacting, even in cases where the observed redshifts for the two galaxies are very different. He suggested that the pair NGC 450/UGC 807 is one of the best cases where the apparently interacting galaxies are well-defined spirals [1,2,3] (see Fig. 1).

NGC 450 is a knotty-armed Sc spiral, viewed at an inclination of 36° ± 8° to the line of sight. Northeast of the nucleus, several giant H II regions are located near the line of sight to the Sb galaxy, UGC 807. The prominence of these knots has been cited by Arp as evidence of the interaction. UGC 807, the companion galaxy, is only one-third as large as NGC 450. At the outset of our program, the systemic velocity of UGC 807 was unknown.

We have shown earlier that the shapes and amplitudes of rotation curves of galaxies can serve as valuable diagnostics of their dynamical conditions. For a normal isolated galaxy, rotational velocities rise fairly rapidly over the first few kiloparsecs from the nucleus, then rise slowly or remain constant across the galaxy. For interacting galaxies, the rotation curves are often abnormal and indicative of tidal disturbances.

Spectra along the major axes of NGC 450 and UGC 807 were obtained with the RC spectrograph plus Carnegie (RCA C33063) two-stage image tube at the Kitt Peak National Observatory 4-meter telescope. The systemic (central) velocity of NGC 450, reduced to the Local Group, is $V_{LG} = 1863 \pm 15$ km sec^{-1}. The systemic velocity of UGC 807 is $V_{LG} = 11583 \pm 50$ km sec^{-1}. Hence if each galaxy is located at the distance implied by its

---

[a] With W. Kent Ford, Jr.

Figure 1. *The nearby spiral NGC 450 partially obscures the more distant galaxy UGC 807. Internal motions show the galaxies to be at very different distances, but superposed along the line-of-sight. Image taken with the Kitt Peak National Observatory 4-meter telescope, and kindly made available by Dr. H. Arp.*

redshift, distances are $D_{450} = 37$ Mpc and $D_{807} = 232$ Mpc, for $H = 50$ km sec^{-1} Mpc^{-1}. UGC 807 would thus be six times as distant as NGC 450, and their association would be no more than a chance appearance along the same line of sight.

Forms of the rotation curves for each galaxy are normal, and show none of the dynamical peculiarities which we have observed in tidally interacting pairs. But we can be even more quantitative, and can compare the rotational properties of each galaxy with those of normal spirals of similar morphology.

We discuss first the companion galaxy. UGC 807 appears to be a normal Sb or Sbc galaxy. The maximum rotational velocity, $V = 211$ km sec^{-1}, identifies UGC 807 as a fairly luminous spiral; its rotation curve appears normal. Its southwestern portion is viewed through the outer regions of NGC 450. When matched against a set of Sb synthetic rotation curves (smoothed descriptions of rotation curves formed from the properties of all the galaxies we have observed), the absolute magnitude of NGC 807 is about $M_B(\text{Sb}) = -21.3$. Matched with the corresponding set of Sc curves, $M_B(\text{Sc}) = -21.8$. We adopt $M_B = -21.6 \pm 0.6$ based on the morphological type and rotational properties. This determination of absolute magnitude (i.e., intrinsic luminosity) is, of course, independent of the distance of the galaxy. Placed at its Hubble distance, UGC 807 would appear as apparent magnitude $m = 15.2$, which is just as observed. If, instead, UGC 807 were located as close as NGC 450, then it should appear enormously bright, $m = 11.2$, which it does not. Or, stated another way, for UGC 807 to be both close and apparently faint, it would have to be a galaxy of extremely low luminosity, $M_B = -17.8$. Its rotational velocity would then be predicted to be less than 100 km sec^{-1}, while we observe 211 km sec^{-1}. Because its rotation curve and indeed its morphology identify UGC 807 as a normal Sb or Sbc galaxy of high luminosity, there is no doubt that it is located at the large distance implied by its systemic velocity.

The brighter galaxy of the pair, NGC 450, has a rotation curve which also appears sufficiently normal as to rule out a tidal interaction. From its shape and low amplitude ($V_{max} = 145$ km sec^{-1}), we assign an absolute magnitude of $M_B = -20.6 \pm 0.7$ when matched to the set of synthetic Sc rotation curves. With $M_B = -20.6$ and a Hubble distance of 37 Mpc, NGC 450 will have an apparent magnitude $m_c = 12.3 \pm 0.7$. This is just as observed. Hence NGC 450 is a moderately low-luminosity Sc galaxy, more than six times closer to us than UGC 807.

The conclusion is clear. The shapes of the rotation curves of NGC 450 and UGC 807 offer no evidence of a tidal interaction. Moreover, when compared with the synthetic rotation curves for Sb and Sc galaxies, the amplitude of the observed rotation curve for each galaxy indicates an absolute magnitude consistent with the observed magnitude only when each galaxy is placed at its Hubble distance. Thus the low-luminosity NGC 450 ($V_{LG} = 1863$ km sec^{-1}, distance = 37 Mpc) is widely separated in space from the higher luminosity UGC 807 ($V_{LG} = 11583$ km sec^{-1}, distance = 232 Mpc). We conclude that this pair is not interacting and hence offers no evidence for noncosmological redshifts.

References

1. H. Arp, Three new cases of galaxies with large discrepant redshifts, Astrophys. J., **239**, 469–474 (1980).
2. H. Arp, Characteristics of companion galaxies, Astrophys. J., **256**, 54–74 (1982).
3. H. Arp, Further examples of companion galaxies with discordant redshifts and their spectral peculiarities, Astrophys. J., **263**, 54–72 (1982).

Some Surprises in M33 and M31

1988

This contribution is a thank you to Frank Kerr for many years of friendship. It stretches the subject matter of this symposium—stretches it enough so that M31 and M33 can be considered far outer parts of our galaxy.

One arcsec subtends 0.04 pc at the center of our galaxy, and 1 arcsec subtends 3.5 pc at M33. But 1 arcsec subtends 100 pc at the Virgo cluster. Thus our resolution from the ground, when we study M31 and M33, exceeds the resolution which will be available in studying Virgo cluster galaxies from the Hubble Space Telescope. For this reason, I have returned to studies of the inner parts of M33 and M31, to see what details we can learn of their morphology, chemistry, and dynamics, using the highest resolution available from the ground with modern detectors.

An Hα image of the central region of M33 shows that the semistellar nucleus is unremarkable, and barely distinguishable from the superposed stars. The near-nuclear region abounds with Hα knots, spheres, and shells, but these do not appear to be organized into any large-scale overall patterns at the light levels displayed in this image.

The chemistry of the nucleus is nearly unique among spiral galaxies, in showing Hα in deep sharp absorption and [NII] λ6548 and λ6583 in emission [1]. However, this uniqueness is related to its proximity. For galaxies at larger distances, a nuclear spectrum contains light from the nucleus as well as light from the disk, so the disk Hα emission blends with the nuclear absorption to produce an Hα emission feature [2,3]. Immediately beyond the nucleus to the Southwest, prominent emission lines of Hα and [NII] comprise a characteristic disk spectrum. However, there is a curious Southwest/Northeast asymmetry. In contrast to the strong disk emission to the Southwest of the nucleus, Northeast of the nucleus the [NII] lines are weak, and Hα shows weakly in absorption. Hence, the spectrum Northeast resembles more a spectrum from the nucleus than a spectrum from the disk. Yet the decrease in surface brightness with distance from the nucleus is symmetric from the Northeast to the Southwest, making it unlikely that nuclear absorption is producing the curiously asymmetric line ratios. This observation gives added evidence to the supposition that nuclei of galaxies are chemically distinct entities, with properties significantly different from their surrounding disks. The asymmetry in line intensity ratios is coupled with a corresponding asymmetry in the velocities of the ionized gas.

It seems likely that M33, an Sc II–III galaxy with minimal bulge, has a shallow central potential in which the relatively low mass nucleus sloshes around. But the asymmetries we observe in M33, both in the dynamics and in the chemistry, are representative of asymmetries which are observed in numerous galaxies including our own, which turn up when the dynamics and morphology are subjected to a detailed look. Another such galaxy is M31.

During a period spanning five years, Ciardullo and his associates regularly obtained CCD images of the near-nuclear regions of M31 in order to search for nova. One filter used covered the [NII] and Hα emission. The frames were taken with the Kitt Peak 36-inch telescope. Ultimately, hundreds of frames were available, from which Ciardullo produced a composite image of the ionized gas near the nucleus of M31 (Fig. 1). The bright spiral covers a region of about 4 arcminutes along the major axis, or about one-thirtieth of the optical disk of M31 [4].

Morphologically, the ionized gas resembles a turbulent spiral with finely patterned filaments and neatly parallel features. Overall, the aspect ratio of the gas appears more face-on than does the outer disk of stars and gas, for which the viewing angle i = 77°. If we approximate the distribution of the gas as a disk, then the Northeast part of the spiral is inclined about 45° to the line-of-sight; the Southwest region is significantly more face on. In particular, the Southwest arm, which extends over 500 pc in length, appears almost circular on the plane of the sky, suggesting an inclination of no more than 20°.

The velocity field of the excited gas near the nucleus of M31 was studied by Rubin and Ford, without the benefit of an image of the nuclear gas [5]. Interestingly, one major feature of the velocity field is a severe asymmetry. On the Northeast, a pattern typical of a rotating disk is observed; the viewing angle suggested by the velocity pattern is near 45°. To the Southwest, the velocity pattern is almost flat, as would be expected in viewing a nearly face-on disk. Hence, both the morphology and the dynamics suggest that the small central gas disk in M31 is not in the princi-

Figure 1. *A negative image in the light of Hα and [NII] of the ionized gas within 8 arcminutes of the center of M31. The image was produced by Ciardullo [4] by superimposing several hundred narrow band CCD frames taken with the Kitt Peak 36-inch telescope. Note the fine filamentary spiral and its relatively face-on appearance, as compared to the aspect ratio of the outer disk of M31. North is to the top, East to the left.*

pal plane of the galaxy, and that this inner gas disk experiences a pronounced twist from the Northeast to the Southwest. The gas layer is very thin, probably less than 80 pc thick. This is indicated by the narrow emission lines.

Overall, it is the asymmetries near the nucleus of M31 which seem most remarkable. Yet asymmetrical features, which are seen on all scales and in all components of M31, have been insufficiently emphasized previously. Dressler and Richstone have shown that M31's kinematical center is displaced 0.5 arcseconds (2 pc) Southwest of the optical nucleus [6]. It has long been known that eccentricities and position angles of M31's elliptical isophotes change rapidly within the first few minutes from the

nucleus [7]. Our observations demonstrate a morphological and kinematical asymmetry in the bulge's ionized gas, and suggest that a warp has
twisted the gas south of the nucleus into a more face-on orientation. On
scale of a kiloparsec, McElroy's observations of the stellar velocity field
shows a gross difference between the velocities Northeast and Southwest
of the nucleus [8]. North of the nucleus, the stars show a rising rotation
curve; south of the nucleus the curve is much flatter. These velocities remarkably mimic those in the ionized gas, but on a larger scale. And on the
largest scales of several kiloparsecs, the HI velocities again show a significant displacement of the kinematic center of the galaxy [9]. While a triaxial
bulge structure in M31 might produce some of these effects, asymmetries
clearly exist also in the disk, where the orbits are believed to be circular.

Studies of the nuclear regions of the nearest spirals offer an opportunity
to learn the details of morphology, chemistry and kinematics that cannot
be obtained for more distant objects. We are continuing the observations
in the hope of solving some of the puzzles we have uncovered.

References

1. V. C. Rubin and W.K. Ford, Jr., Ap. J. Letters, **305**, L35 (1986).
2. W. C. Keel, Ap. J., **268**, 632 (1983).
3. A. Filippenko and W. L. W. Sargent, Ap. J. Suppl., **57**, 503 (1985).
4. R. Ciardullo, V. C. Rubin, G. H. Jacoby, H. Ford, and W. K. Ford, Jr., A. J. **95**, 438 (1988).
5. V. C. Rubin and W. K. Ford, Jr., Ap. J., **170**, 25 (1971).
6. A. Dressler and D. O. Richstone, Ap. J. **324**, 701 (1988).
7. S. A. Kent, Ap. J., **266**, 562 (1983).
8. D. B. McElroy, Ap. J., **270**, 485 (1983).
9. E. Brinks, Ph.D. thesis, Rijksuniversiteit te Leiden (1984).

NGC 4550: A Two-Way Galaxy

1993

N GC 4550 is a normal-appearing galaxy located near the center of the Virgo cluster. It is classified as E7/S0, where the E7 denotes a highly flattened featureless elliptical, and the S0 indicates the presence of a distinct disk (Fig. 1). It has one unique, important characteristic: its disk plane contains stars about evenly divided between stars that orbit the center of NGC 4550 in a clockwise direction, and stars that orbit counterclockwise. Before this discovery, astronomers thought that all galaxy disks contained stars orbiting in essentially the same direction. But now we know that at least one bidirectional galaxy exists, and we understand in a general way how it came into being. This note describes the discovery and the work that was necessary before I was convinced that I understood the complexity within NGC 4550.

In a normal galaxy disk, stars move in circular orbits about the galactic center with velocities in the range of 80–320 kilometers/second (50–200 miles/second). Our sun orbits the center of our galaxy with a velocity of about 220 km/second (140 miles/second). When we view an inclined galaxy, stars on these orbits move toward the observer on one side of the disk, and away from the observer on the other (with respect to the velocity of the galaxy center). Hence, we observe in the spectrum of a typical disk galaxy emission lines (from the gas) and absorption lines (from the stars) in a pattern that is Doppler shifted to bluer wavelengths on one side of the disk (the approaching side) and to redder wavelengths on the other (receding) side.

Several years ago I embarked on an extensive observing program, in collaboration with Jeffrey Kenney of Yale University, to study the rotation properties of stars and gas in 80 galaxies in the Virgo cluster. As part of the project, I took our first spectrum of NGC 4550 in 1989 at the Palomar 200-inch telescope using a spectrograph with an electronic CCD detector; the

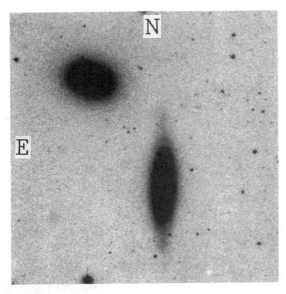

Figure 1. *The galaxy pair, NGC 4550 (bottom) and 4551 (top), observed by C. Bailyn and Y.-C. Kim at the Cerro Tololo Inter-American Observatory. Although stars in NGC 4550 are located in a disk in which half of the stars orbit clockwise, and half of the stars orbit counterclockwise, the galaxy shows no morphological peculiarity.*

exposure was 2,000 sec, with the spectrograph slit placed along the galaxy disk. A second spectrum was obtained in 1990 at Kitt Peak National Observatory's 4-meter telescope, with the slit placed perpendicular to the disk, along the minor axis.

Study of the two spectra showed immediately that NGC 4550 is a rare object. At the location of the Hα line of hydrogen, the region where the spectra are centered, the absorption lines show the normal pattern of rotation from the stars (Fig. 2a) but also a hydrogen emission line from a *gas* disk which is rotating in the direction opposite to the stellar disk (Fig. 2b). About a dozen galaxies containing these so-called counter-rotating gas disks are known. But as I continued to study the spectra, I had the vague feeling that I was detecting a pattern more complex than one stellar disk and one oppositely rotating gas disk. On my spectra, always optimized for detecting emission lines, stellar absorption lines are weak. In fact, astronomers who routinely derive motions from absorption lines do not measure the spectra visually, line-by-line (as I measure the emission lines). Instead, they use complex computer algorithms to extract line shape information from all the weak lines in order to synthesize a single absorption line pattern.

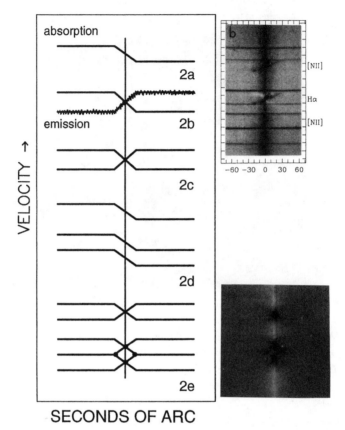

Figure 2. (a) An absorption line from starlight in a galaxy disk, sketched as it would appear in an astronomer's spectrum. The vertical axis charts increasing wavelength and thus velocity. Light from stars in orbit about the center of the galaxy is shifted to higher wavelengths on the side of the galaxy where orbits carry the stars away from the Earth. The greater the velocity, the higher the shift. The horizontal axis represents distance from the center of the galaxy, measured from the long vertical line. (b) Left: Sketch of an absorption line and an emission line from a counter-rotating gas disk. Right: A negative print of a portion of the spectrum NGC 4550, showing the Hα absorption line from the stars (light), and an Hα emission line (dark) from the counter-rotating gas. The straight lines crossing the spectrum are emission from the Earth's atmosphere. (c) Sketch of a spectrum with two counter-rotating disks. (d) Sketch of three absorption lines, called a triplet, due to magnetism. (e) Left: Sketch of the Mg triplet, with bidirectional disks. Note the diamond pattern where the lines intersect. Right: A positive print showing the Mg triplet arising from disk stars in NGC 4550. The pattern formed by the intersection of the absorption lines from stars on clockwise and counter-clockwise orbits matches that predicted by the sketch.

For months, I would alternately believe and then not believe that I could see a normal spectrum composed of many weak absorption lines, but that each absorption line was double, with the two members opposite in velocity and crossing at the nucleus (Fig. 2c). When my colleagues John Graham and François Schweizer passed by the image processing terminal, they too were roped into the "Do you believe it?" game. Ultimately, a curious pattern persuaded me that every line was double. Three lines of magnesium, normally a singlet next to a doublet (Fig. 2d), showed a puzzling diamond pattern unlike any I had ever seen. In an effort to understand what I was looking at, I sketched what these three lines would look like if each were double, with a corresponding member from the opposite disk. And there, clearly, where the doublet lines intersected, was reproduced the diamond pattern (Fig. 2e). That was an exciting moment.

Meanwhile I collected theoretical studies relating to bidirectional disks from the astronomical literature. These patterns, permitted by the laws of physics, had been assumed to be unrealizable in nature because no method was known for establishing such a disk. Donald Lynden-Bell had discussed them first, in a paper I read in 1961 and had never forgotten. He employed a "Maxwell demon" (an imaginary creature) to reverse the direction of half the stars in a spherical cluster. He showed that the resulting system of two interpenetrating and counter-rotating spherical clusters is stable. Later, Alar Toomre discussed bidirectional disks as "elegant curiosities." James Binney and Scott Tremaine in their outstanding *Galactic Dynamics* text dismissed such models with the footnote, "However, it is unlikely that this type of model is of practical importance." I await the next edition.

How can we understand this remarkable galaxy? It is hard to envision a single formation event, which, out of a gaseous protogalaxy, would produce stars orbiting in two opposite directions. For example, in our solar system, (which is disk-like but 100 million times smaller than a galaxy), the fact that all the planets orbit the sun in the same direction is used to infer that the planets formed from the same circumstellar disk of gas and dust. Hence NGC 4550 must have undergone a "second event," a capture of external gas whose orbital direction was opposite to that of the existing galaxy.

Computer simulations show that captured gas will be funneled toward the nucleus, and complex velocity patterns can emerge. After the gas has settled to the plane of the existing galaxy, stars form, creating a second stellar disk. We think that the merger began at least several billion years ago. The ten or more orbital periods since the cataclysm have permitted stars to form, evolve, and produce metals, and have erased all morphological peculiarities. In a disk galaxy, stars are very far apart relative to their diameters. In fact, it is possible to line up 10^8–10^9 stars, i.e., 1% of the 100 billion stars in a galaxy, between the average separation of two stars. Hence, two oppositely rotating stellar disks will interpenetrate with little interference once such a system has been established (as will also a stellar disk with a counter-rotating gas disk).

NGC 4550 is so far unique in exhibiting two counter-rotating, extended stellar disks. The few rare galaxies with a stellar disk and a counter-rotating gas disk are those which have not yet started turning their newly acquired gas into stars. And the even fewer galaxies with a stellar disk and a small nuclear counter-rotating *stellar core* must be in a still later evolutionary phase.

There is a coda to this discovery path. In March 1992, John Graham and I were able to get a confirming spectrum at the Kitt Peak four-meter telescope; only then did we submit a paper for publication. And in April 1992, I discussed NGC 4550 at a Harvard University astronomy colloquium. Marijn Franx, a member of the audience, informed me that he and his colleagues, Hans-Walter Rix, D. Barnes, and Garth Illingworth, experts in the analysis of absorption-line galaxies, had existing spectra of NGC 4550, but had not noticed its peculiarities. Within days, their re-analysis of the spectra confirmed the double nature of the lines. Later came a note from Scott Tremaine: "bidirectional disks are more stable than unidirectional disks." Still later, Paul Schechter of MIT reminded me that long ago I had speculated to him on the existence of galaxies with a counter-rotating disk of stars. I anticipate the discovery of more, but probably not many more.

This extended path to understanding the stellar kinematics within NGC 4550 is hardly the flash of insight we often read about. Yet it is enormously satisfying to have the eye of the observer rewarded occasionally with a view of a former secret of the universe. But I envy the stellar observers *in* NGC 4550. They must have figured it out long before we did.

Part II
TOOLS OF THE TRADE:
TELESCOPES, A CATALOG,
AND SOME MAPS

A grating, a spectrograph, and a telescope. For a few nights I use these instruments at a remote mountain site to gather data that I carry back to my office to analyze over many months. Telescope time is precious. Applications to use the U.S. national facilities at Kitt Peak Observatory are often oversubscribed by factors of 3 or 4. The truly innovative part of designing an observing program involves intuition and guesses. What is the new (novel?) question to be asked of nature; what observing strategy is most likely to lead to a conclusive answer; what specific objects are promising candidates for study; what observing techniques will exploit the capabilities of the instrumentation? It is exhilarating to watch answers accrue.

Atlases, star charts, and globes, once the domain of the artist/scientist, remain a beautiful part of astronomy. My interest in celestial globes began when Maud Makemson, my professor of Astronomy at Vassar College, offered me an old celestial globe about to be discarded. A dedicated teacher, she may have known then that she was opening many doors for me; it was some years before I knew it.

"Astronomy from Hubble" is an after dinner talk presented to the Visiting Committee of the Department of Terrestrial Magnetism, Carnegie Institution of Washington, 1994.

"Mount Wilson Observatory: A Brief Early History" was spoken at a congressional breakfast hosted in 1992 by the Honorable George E. Brown, Congressman from the Mount Wilson, California district.

"On the Dedication of the Vatican Advanced Technology Telescope" was delivered in Tucson 1993, at the Inaugural Dinner of the Vatican Telescope.

"Letter from Chile" was written in Santiago Chile in 1971, following observing at Cerro Tololo Inter-American Observatory and the Carnegie Las Campanas Observatory. I was awaiting the arrival of my husband for his first Chile visit. The letter remained unpublished in my desk drawer until now.

"Observing at the National Facilities" appeared in A Look at AURA 1994. AURA is the Association of Universities for Research in Astronomy, which operates the Kitt Peak, Cerro Tololo, Sacramento Peak Observatories, and the Space Telescope Science Institute, under agreement with the National Science Foundation. These U.S. national facilities are open to all astronomers through competitive access.

"A Revised Shapley–Ames Catalog of Bright Galaxies" was a 1982 review published in Sky and Telescope. The Shapley–Ames catalog continues to play the role of a dictionary for observers of galaxies. In 1994 Allan Sandage and John Bedke published a massive two volume Carnegie Atlas of Galaxies (Carnegie Institution of Washington, 1994), to accompany and supplement the earlier revision. The new atlas, containing exquisite images of 1,225 galaxies, is the pinnacle of photographic extragalactic astronomy and will probably never again be equaled, as astronomers turn from the fine-grained, wide-field photographic plate to the presently coarser and smaller digital detector.

"Star Charts," a review of The Sky Explored, by Deborah J. Warner, was published in Science, 1980.

Astronomy from Hubble

1994

When we look with a ground-based telescope at the Andromeda galaxy, the nearest spiral galaxy, we see a central bulge—the light from billions of stars combined by the earth's atmosphere into one large blob—surrounded by spiral arms defined by bright knots of newly formed stars. Under conditions of perfect optics and no atmosphere, the stars in the bulge would not be blended, but would appear as distinct individual points. Over most of the bulge of M31, the stars are actually far apart, covering only 1/40,000 of the available surface area. In an effort to produce better images than those possible below the atmosphere, astronomers designed the Hubble Space Telescope to fly above the Earth. The lack of atmospheric distortion increases the resolution by factors up to five or more. This is a major reason why astronomers send telescopes into space.

After decades of study, design, and production, the Hubble Space Telescope was launched in the spring of 1990. On June 21, 1990, NASA announced that an incorrectly spaced field lens used in the manufacture of the primary mirror had resulted in a spherical aberration, an aberration astronomers recognized with dismay following the first images.

In terms of lens design the error is so large that simple testing like that routinely done by amateur telescope builders would have detected it. In absolute measure, the error is small: the media reported an error "less than 1/50th the thickness of a human hair." In addition to the anguish and disappointment within the scientific community, many of us had to live down the indignity of a cartoon at every colloquium; every failure was compared to the HST. Even the friendly *Sky and Telescope* magazine embarrassed astronomers with its cartoons.

But optics is an exact science, and eyeglasses could be prescribed to make the viewing essentially as originally designed. In a dramatic repair mission in December 1993 the wide-field camera was replaced by a camera with corrected optics; and corrective lenses were placed to intercept the light en route to the other instruments. And now astronomers, and even speed skaters, shine in the reflected glory. In reporting Dan Jansen's 1000-meter victory in Olympic speed skating in February 1994, Tony Kornheiser wrote in the *Washington Post*, "So *this* is what it feels like. You could see the relief in his eyes. They've always been bright and blue, but this time they were clearer than the fancy new mirrors of the Hubble Telescope."

The corrected images are close to perfect; we now have a great telescope. But any telescope, even the unique and spectacular HST, is still just a tool. As in all science, the important decision is what programs we choose to study with it. There are as many possibilities as astronomers: at the Department of Terrestrial Magnetism we have the real luxury of choice. It is a mistake to think that science is only curiosity-driven. It is driven too by what we have learned in the past, and by what we perceive are the directions that will lead to more learning. With enthusiasm and hope, we make our most educated guesses as to what new observations will permit nature to reveal one more of her secrets.

We live in a solar system dirtied by the debris of its formation. Comets are one manifestation of these leftovers. Of the few images already released from the corrected Hubble, the picture of comet Shoemaker–Levy is exceptionally dramatic. Discovered from the ground about a year ago by Caroline and Gene Shoemaker and David Levy, the comet had previously passed by Jupiter in summer, 1992, where it had been broken up into 9 or 10 pieces by Jupiter's gravitational field. Imaging with HST has identified 20 or so fragments, some in the process of appearing, some in the process of disappearing. The comet's orbit will carry it back to the planet Jupiter in July 1994, where each piece of the comet will successively impact the planet. Sadly, the impact will take place on the far side, but clever observations will be carried out in an effort to detect evidence of the splashes. David Rabinowitz, one of our postdocs and an expert in detecting solar system debris, will use the Carnegie 100-inch telescope at Las Campanas to watch the Jovian moons, to see if light from the impacts is reflected to earth by those moons.

A major program of the HST is to determine accurate distances to nearby galaxies. For the nearest galaxies, this means identifying Cepheid variable stars and mapping their light variation. Cepheids are stars in advanced stages of evolution whose outer layers are temporally unstable, causing the star to pulsate. Cepheids are invaluable as distance indicators, for their light variation is proportional to their intrinsic brightness; a low luminosity Cepheid will go from faint to bright to faint in a day; a high luminosity

Cepheid can take 100 days. Hence by establishing its period of brightening, the true brightness of the star can be determined and then compared with its apparent brightness to deduce the distance to the galaxy.

Unfortunately, even for the nearest galaxies observed from the ground, a single resolution element contains several stars, so the light variation of a single star is diluted. Many hours of observations with the 200-inch telescope from Palomar had resulted in the discovery of only 2 Cepheids in M81, a nearby spiral galaxy. A group of astronomers observing M81 with the HST have obtained 12 images in M81, from which they have found 37 Cepheids, and determined periods for 27. It is anticipated that HST will determine accurate distances to galaxies as far as the Virgo supercluster. But the next step, the determination of the rate of expansion of the universe, will not come easily. Galaxies have random motions, and groups of galaxies stream toward regions of higher galaxy density. Both of these effects will mask the rate of the large-scale expansion. Such motions, complex and difficult to calculate accurately, cause major complication in the evaluation of the Hubble constant.

For more distant galaxies, HST will reveal previously undetected features. François Schweizer has devoted much of his professional career to the study of merging galaxies, and the evolutionary effects therefrom. One of his favorite galaxies is NGC 7252, sometimes called the Atoms-for-Peace Galaxy because of the complex wrapping of its tidal tails. Schweizer has shown that NGC 7252 is on its way to becoming a spheroidal galaxy, the result of the merger of two disk galaxies. In an effort to confirm his suspicion that globular clusters (massive accumulations of hundreds of thousands of stars) are also formed during a merger, Schweizer and Brad Whitmore have observed NGC 7252 with HST, where the high spatial resolution permits the separation of non-point globular clusters from point (foreground) stars.

And as the observers had hoped, they discovered some two dozen young globular clusters encircling the nucleus of NGC 7252. Indeed, young globular clusters are produced in a merger. But the observers had another spectacular reward as well: a view of a central mini-spiral, tiny yet precise (Fig. 1) whose existence had never been suspected. Its formation can be understood as a by-product of the gas funnelled to the galaxy center by the gravitational interaction. Hence both the clusters and the spiral are young and are products of the merger. Because of its youth, the brightest cluster is 600 times as bright as the much older brightest globular cluster in our Milky Way Galaxy.

So when we catalogue the individual parts of NGC 7252: a central disk from one galaxy, rotating skew with respect to a disk from its second,

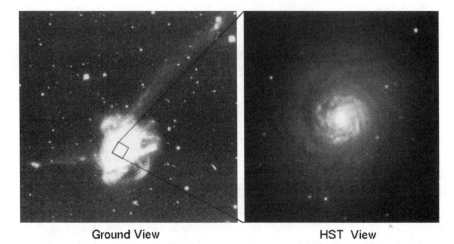

Ground View   HST View

Figure 1. NGC 7252, (left) reproduced from a plate taken by François Schweizer
with the 4-meter telescope of the Cerro Tololo Inter-American Observatory; exposure 85 min. The prominent outer loops and tails are diagnostic of a merger.
The center box marks the region shown in the right image. The nuclear region
(right) of NGC 7252, from an image taken with the Hubble Space Telescope,
reveals a "mini-spiral" of gas and stars, plus many bright, young globular star
clusters.

merged galaxy; active central star formation; young globular clusters;
infalling tidal tails; we are incredibly viewing pieces both of its past history and of its path toward becoming a future spheroidal galaxy. These
observations, coupled with previous studies, offer strong evidence that
galaxies are assembled piece by piece by merger processes, as surely as
California was assembled piece by piece by geologic tectonic processes.
Just as the Earth is evolving beneath our feet, so too are galaxies evolving
over our heads. This view of galaxy evolution is a major change since
Edwin Hubble was a Carnegie Institution of Washington employee. Hubble
envisioned galaxies as island universes evolving in splendid isolation, products only of their heredity. The major role of environment was unimagined.
But other things have not altered since Hubble's era. When asked what he
expected to find with the new 200-inch telescope, Hubble replied, "We
hope to find something we hadn't expected." François is lucky: he has
already found something he hadn't expected.

Mount Wilson Observatory: A Brief Early History

1992

The twentieth century will be remembered as the century in which we learned what a galaxy is. Work carried out at the Mount Wilson Observatory played a major role in that learning process. This morning I will describe in a very few minutes and in a very casual manner the importance of two great men, Andrew Carnegie and George E. Hale, to the early history of this Observatory, and some of the early science which followed.

1902: Andrew Carnegie discussed forming the Carnegie Institution of Washington; Langley, the Secretary of the Smithsonian, recommended a solar observatory at a high mountain site. A site survey in southern California recommended Mount Wilson.

1903: Hale wrote to Carnegie of his interest in a solar observatory. Hale, at the Yerkes Observatory, had earlier been highly influential in getting the 40-inch refractor built at Yerkes; it was at that time the largest telescope in the world.

1904: Carnegie authorized a grant to found the Mount Wilson Solar Observatory, with Hale as Director. And in what surely must have been a first (and a last?) in the history of astronomy, Hale arrived in Pasadena with a 60-inch glass disk, which his father had bought for him in 1896, but which the Yerkes Observatory could not afford to mount.

1908–1917: After a series of trials and set-backs, including the San Francisco earthquake of 1906 which nearly destroyed the telescope mount being constructed there; a fire on Mt. Wilson which damaged the observatory site; and the problems of transporting telescope parts by mules up a poor mountain road, the mirror was finally installed and the telescope initiated in 1908. The largest telescope in the world was now located on

Figure 1. Andrew Carnegie and George Ellery Hale in front of the 60-inch tele-scope, Mount Wilson, March 21, 1910. (Photograph courtesy of Carnegie Insti-tution of Washington.)

Mt. Wilson. When Carnegie visited the site for the first time the newspa-pers headlined the next day "Rain Shuts Out Stars He Hoped to See" (Fig. 1). But Hale, never one to rest on past achievements, had already begun planning back in 1906 for an even larger telescope, the Mt. Wilson 100-inch. Made at the Saint Gobain factory in France, the blank arrived in California in 1908, grinding was started in 1910, and the telescope was completed in November, 1917. The Mount Wilson Solar Observatory be-came the Mount Wilson Observatory.

The young astronomer Harlow Shapley arrived at Mt. Wilson in 1914, and spent the next five years studying the 100 or so known globular clusters, each containing hundreds of thousands of stars. In the course of this work, he made the bold assumption that their curious distribution on the sky resulted from their placement in a spherical halo about our galaxy. Moreover, he contended that the geometrical center of their distribution, in the dense star clouds in the southern Milky Way in the constellation of Sagittarius, defined the center of our galaxy. Thus, he taught us, *the center of the galaxy is not located at the sun's position*. Just as Copernicus several centuries earlier had displaced the Earth from the center of the solar system, Shapley now displaced the sun from the center of the galaxy. This was a profound achievement in the history of astronomy, but more followed.

In 1919 Edwin Hubble arrived at Mount Wilson to use the 60-inch and the 100-inch telescopes in order to study the "island universes," systems of billions of stars which he identified as individual galaxies, like our own Milky Way Galaxy, but located at vast distances. His studies were the first to show convincingly that the universe is composed of galaxies which are separating one from the other. From his work, we learned that the universe is indeed expanding, as was predicted earlier by Einstein's cosmologies.

This, then, is some of the knowledge of the universe which we inherited mid-century, and which remains the foundation for the intense worldwide work in extragalactic astronomy that goes on today. In future efforts, the Observatories of the Carnegie Institution of Washington will surely continue to play a major role.

The Dedication of the Vatican Telescope

1993

I t is always a pleasure to celebrate a new telescope with which to view the heavens. But of course what we also celebrate tonight is the wonder and the glory of the universe. We live in a remarkable universe: amazingly beautiful, enormously large, incredibly complex. As youngsters, all of us are curious about the sky, the sun, the moon, the stars. Some of us never lose this curiosity, and we are privileged to devote a major effort of our lives to studying and attempting to understand the universe. We are called astronomers.

A telescope is our tool, and this telescope is a very special one, as it initiates a new generation of telescopes. It has a new type of mirror that is polished in a new manner; it is mounted in a different way, and it will be used by a unique set of astronomers—astronomers of the Vatican Observatory. There have been astronomers at the Vatican for over four centuries, and the present Vatican Observatory was founded in 1891 by Pope Leo XIII. The distinguished line of Vatican astronomers includes the beloved Fr. Secchi, who very early founded the science of stellar composition.

But the most important feature of any telescope is the imagination with which it is used. It has been said of Galileo that he advanced astronomy when he took a long tube and put a small lens at one end and a large brain at the other.

It is not possible to predict the many discoveries which will be made with the Alice Lennon Telescope and the Thomas J. Bannan Astrophysics Facility. I would like to tell you, however, what we *do* presently know about the universe, for this is the knowledge which will form the basis of the observing programs carried out with this telescope.

We live *on* a small planet, which orbits an average star; it takes one year for our planet to orbit around the sun. And the sun, plus all of the stars in our galaxy, orbit in concert about a distant center. Our sun is one of the 200 billion stars in our galaxy; it takes 200 million years for the sun to orbit once about the center of the galaxy. So we, living on spaceship Earth and orbiting about the sun, are carried by the sun at a speed of one-half million miles per hour about the center of the galaxy. Don't ever let anyone tell you that you haven't been on a spaceship. During the two hours we have been celebrating, we have traveled one million miles.

We live *in* a spiral galaxy. This is a wonderful thing to know. When you look at the sky on a dark night from the Northern Hemisphere, all of the stars that you can see with the naked eye belong to our galaxy. On a very very clear night, a faint fuzz can also be seen; the Andromeda galaxy. And when you stand on a dark moonless night at a dark site and see the glorious Milky Way, you are seeing the stars in the disk of our galaxy.

During the 20th century, astronomers learned that the universe is filled with galaxies, that the universe is expanding and galaxies are separating one from the other. But perhaps more important than the individual facts is our new understanding that the universe is more complex and more mysterious than we had ever previously imagined. Never will this telescope run out of things to study. Diane Ackerman, who eloquently describes nature, has recently written:

> It is nighttime on the planet Earth. But that is only a whim of nature, a result of our planet rolling in space at 1,000 miles per minute. What we call "night" is the time we spend facing the secret reaches of space, where other solar systems and, perhaps, other planetarians dwell. Don't think of night as the absence of day; think of it as a kind of freedom. Turned away from our sun, we see the dawning of far-flung galaxies. We are no longer sun-blind to the star-coated universe we inhabit.

So let us take a very rapid tour out to the limits of what we can see with our telescopes, to distances so far that it takes light 10 billion years to reach us.

Figure 1 is a schematic diagram of our universe, at least that part of it that we study with conventional telescopes. We live on a planet orbiting the sun; it takes 8 minutes for light to reach us from the sun. Our sun is one of hundreds of billions of stars in our galaxy, all orbiting about a distant center. It takes 30,000 years for light from the center of the galaxy to reach us. Stars in a galaxy are very far apart, but galaxies are very close together relative to their diameters. We have 2 satellite galaxies, the Large and Small Magellanic Clouds which are orbiting our galaxy. They, plus the Andromeda galaxy, plus its satellites, plus about 2 dozen small, faint conglomerations of stars make up the Local Group, our local region of the universe.

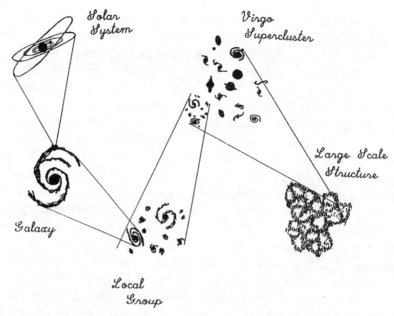

Figure 1. *A sketch of the observed luminous universe, emphasizing the lumpy structure. We live on a planet orbiting a star; the star is located in a spiral arm of our Milky Way Galaxy, which is one galaxy in the Local Group of galaxies. The Local Group is an outlying member of the Virgo Supercluster, which forms one of the knots in the lumpy universe.*

The Local Group is a suburban member of the Virgo Supercluster; it takes light 40 million years to reach us from the Virgo Supercluster. And the Virgo Supercluster is just one of the lumps in the perhaps lacy, perhaps sudsy, large-scale structure which we see at the limits of our observations to date.

The lumpy large-scale distribution of galaxies is a puzzle to some astronomers, especially theorists. Present day galaxies are presumed to be relics of lumps in the early universe, yet the early universe exhibits very little evidence for significant irregularities. Thus it is difficult to relate the lumpy observed distribution to the early smoothness.

That's what we have learned. But the Vatican telescope will be looking often for what we have not yet learned. And what better words of wisdom could be found with which to set out on this journey than these of Proust, *"The real voyage of discovery consists not in seeking new landscapes but in having new eyes."* Real astronomical discoveries come from looking with new eyes and in asking new questions.

Some questions to be attacked with the new telescope:

- How do stars form?
- What of the epoch before there were galaxies?: What preceded the first generation of stars; perhaps gas clouds?
- What is dark matter (hard one for optical telescope)?; is it conventional or exotic?
- Is there life elsewhere? How can we detect signs of other life in the universe?
- How much matter is in the universe? Do we live in a low-density universe that will expand forever, or in a high-density universe which will ultimately collapse?
- Is ours only one of many universes; are there others too distant to be observed?

One night, one-hundred-and-forty years ago, Ralph Waldo Emerson visited the observatory of Williams College, currently the oldest continuously operating astronomical observatory in America, and wrote in his journal:

> I saw tonight in the observatory, through Alvan Clark's telescope, the Dumb-Bell nebula in the Fox and Geese Constellation, the four double stars in Lyra; the double stars of Castor; the two hundred stars of the Pleiades ... I have rarely been so much gratified. Of all tools, an observatory is the most sublime ... The sublime attaches to the door and to the first stair as you ascend;—that this is the road to the stars. Every fixture and instrument in the building, every nail and pin, has a direct reference to the Milky Way, the fixed stars, and the nebulae.

Sublime: What a lovely and true word to describe a telescope! I like the word because it conveys notions of awe, of wonder, of majesty, of grandeur; words which also describe the universe, and will also describe the important discoveries to be made with the Vatican Advanced Technology Telescope. The words awe, wonder, majesty, are important to the imaginative use of the telescope, as are the words advanced technology. All together, they define this telescope, and the work it is meant to do.

And now we all join in wishing the Vatican Advanced Technology Telescope, its designers, its builders, its supporters, and its users, much joy and success throughout many decades of use. May all, looking with new eyes, contribute to expanding our knowledge of this wondrous universe.

Letter from Chile

August 1971

The bus ride from Santiago to La Serena, 300 miles north, is a continuous series of contradictions, as is much of Chile, to the eye of one from the eastern United States. The smog of Santiago hides the snow-covered peaks of the Cordillera, but is gone after 15 miles, and the spectacular view to the east is reminiscent of much of Switzerland: a high continuous snow-covered range in the distance, with green hills and pleasant valleys nearby. Winter here in August, the carnations are even now starting to bloom, and the large fields of flower farms are speckled faintly with color. Last October, along the same route, the spring rains had turned the countryside into a continuous garden of wild flowers. A few hundred meters to the west, a wild Pacific Ocean attacks the rocky coves along the shore. No smog here, but another sign of pollution in the scum which churns continuously off-shore.

Along the Panamericana highway, the countryside is virtually deserted, except for the flower farms near Santiago, or the occasional cardboard hut or stone house of a poor farmer. Most of the land appears unoccupied except by the few houses and cactus. Even the political posters are far apart: "Land is the bread of Chile." Closer to La Serena, the scene is still more desert-like, reminiscent of arid Utah. Large patches of snow dot the southern slopes of the distant hills, an anomaly to the northern eye 2000 miles south of the equator.

The highway deteriorates, the result of a recent earthquake, as we near the quaint artisan town of La Ligua. Here, skilled weavers produce the fabric and ponchos they sell to the tourists. The economic problems of Chile cannot be avoided by the visitor. Travelers' checks are accepted, but only undated; the rampant inflation makes a future exchange rate more favorable. Gasoline is difficult to find, long lines, waits, and negotiations are usual.

La Serena, the first destination, is a thriving town of painted stucco homes, and the bright colors in the sunlight bring to mind an old Italian town. From La Serena, the road hugs the broad Elqui river with its luxurious vineyards and orchards. In the summer, apricots drying on the roofs will give evidence of a good harvest. Less than an hour inland, we start the climb to the observatory atop Cerro Tololo. Large cacti cover the otherwise barren hillsides. Occasional views of the distant telescopes are connected by the colorful shadings in the mineral-rich mountains.

For the astronomer, the night begins with an early supper, for by 5:30 PM I can be at the telescope, setting up the equipment for the night's program. The sun will set within the hour, and will rise 12 hours later. During that time of darkness, each minute will be precious, and I will be aware that I am racing the sun. I arrive with a motley collection; heavy socks, underwear, flashlight, thermos. Closed during the day to keep out the warm air, the telescope dome is now opened to allow the telescope to reach the temperature of the cool night air. It is already 40° in the never-heated dome, and the lights are out, as they will be all night, to protect the dark adaptation of the observers.

I am joined in the dome by a Chilean night assistant, one of a group of bright, enthusiastic, nearly bilingual aides who operate the telescope and help with guiding during the long exposures, if the observer wishes. The assistants at Tololo are an unusually able group, and they expand their duties to include entertaining the observer on a cloudy night with games of Scrabble, either in Spanish or in English.

As darkness falls, the telescope is set on a bright star for focusing and for making small corrections to the coordinates as read on the barely lit console. The rest of the night, I will be photographing on glass plates spectra of galaxies so distant that they appear small and faint and are difficult to detect against the dark night sky. The telescope is set from very accurate positions; a photograph of that sky region taken earlier is examined under a dim red light, and the view through the telescope is searched to find the object. Tonight, I am attempting to get velocities of galaxies in order to study details of the expansion of the universe.

For the first exposure, I take about 15 minutes to locate the faint galaxy and center it exactly, and then locate a nearby bright star on which I can guide the telescope. During the next 90 minutes I stand at the telescope, too intent to sit, keeping the crosswires fixed exactly on the bright star in the viewing scope. About once every minute I correct slightly as the star starts to drift from the crosswires. At Kitt Peak National Observatory out-

side of Tucson, Arizona, the sister observatory to Cerro Tololo, an electronic device guides the telescope during the exposure, but such equipment is not yet available in Chile. I enjoy both modes of observing. Guiding is fun, and standing in the cold dome throughout the night makes the
science extremely personal. Especially during my long six-hour exposures
of Andromeda a few years earlier, I would sit as I guided on the nucleus of
that galaxy and wonder if someone there was sitting in a cold dome looking at our Milky Way. Often I wished we could exchange views.

The 12-hour night is used up with six exposures. I vary times slightly as
the brightness of the object requires, and modify the proposed program
slightly as the science requires. For one exposure, a bright star in the field
was suspected of being a supernova, so an exposure was taken of the star.
Only a star, no supernova. The night is cold, the work is tedious, and the
long hours pass slowly. My vigil is interrupted occasionally when I leave
the dome to develop a plate, also in total darkness, to confirm that all is
going well. During these minutes, the night assistant takes over the guiding.

Morning brings the end of observing for the night, and brings the sun,
and the spectacular view of the coastal fogs filling in all the valleys between the high peaks. With the sun my circadian rhythm is reset and I am
wide awake, too wide awake to sleep and too curious to leave the night's
plates undeveloped, so I walk down to the dark room. Even though the
dark reminds me of my tired body, I know that it is futile to attempt to
sleep while wondering what is captured on the photographic plates. So I
remind myself of my advice to my postdocs, "Observational astronomy is
still an art, the art of making as few mistakes as possible."

One mistake I try never to make is to incorrectly set the darkroom timer;
I must be sure that its buzz will really mark two minutes and I will take the
plates out of the developer at the proper moment. To be certain of my
timing, I sing along my 'two-minute song,' "In the Still of the Night." In
about an hour, the plates are all developed, rinsed, and set out to dry.

I rejoin the bright mountains and walk down the road to the dormitory.
The muted colors on the mountains, the snow on the peaks, the bright blue
sky, and the joy of the night bring to mind the first few lines of the "Panamanian Tamborito Dance" of Gabriela Mistral, herself a native of the Elqui
Valley.

De una parte mar de espojos,	To one side, a sea of mirrors
de la otra serranía,	To the other mountain ranges,
y partiéndonos la noche	And cutting across the night
el tambor de la alegria.	The drum of happiness.[a]

---

[a] "Panamanian Tamborito Dance" by Gabriela Mistral [translation from Selected Poems:
Trans. Langston Hughes (Bloomington–Indiana U. Press, 1959)].

Reminiscences 1994: Observing at the National Facilities

1994

I am an astronomer because the beauty and the mystery of the night sky captivated me as a youngster. I am an *observational* astronomer because the National Observatories offer me continued access to state-of-the-art telescopes, and they permit me to pursue programs whose results reveal a few of the secrets of the universe, and also teach us how much more there is to learn.

My career grew up with the national facilities. In the spring of 1963, the American Astronomical Society met in Tucson, and I toured Kitt Peak with the group. The #1 36-inch telescope, the largest operating telescope on the mountain, displayed its embossed Dymo labels: "Pull here to see star." "Push here to expose comparison." As a young astronomer with only student observing experience, my response was instantaneous: "I could do that." A few months later I was at the 36-inch, observing stars at the outer limits of our galaxy, to learn how they orbit, and to deduce the mass of the galaxy.

By 1965, I had progressed to the new 84-inch telescope, joined the Carnegie Institution and was working with Kent Ford using an image-tube spectrograph to study orbital motions, but now in the Andromeda galaxy, the nearest large spiral to us. Even with a fast image-tube spectrograph and photographic plates, exposure times were long. Often, I would split the cold 12-hour night into two 6-hour exposures, my eye at the guiding eyepiece throughout the night. Telescope operators hated those nights, for they were required to sit at the dome console with nothing to do for 12 hours once they had set the telescope on the galaxy. Because we feared light leaks within the spectrograph, we observed in total darkness; when they saw me coming, the telescope operators quickly covered even the luminous face of the alarm clock.

My first view of the modern world of observing came one evening when Roger Lynds offered me an automatic guider for the night, and the next morning I detected individual features of unusual clarity on the galaxy spectra. This from an observer who prided herself on her guiding ability! But more was to come. In the early 1970s, the 4-meter Kitt Peak image-tube spectrograph came on line, almost totally computer-controlled. The observer sat in the warm console room, bored, until the exposure was done. Then moments of frantic activity followed while the next object was acquired and the exposure started, leaving time to run to the darkroom to develop the previous plate. My best telescopic views of galaxies came from the early 4-meter observations, when it was still necessary to work at the telescope to center the galaxy on the spectrograph slit.

Sometimes the observing run was in Chile, first at the 60-inch, then the 4-meter. Telescope operators there still remember the procedure of rotating the spectrograph at the telescope in total darkness, by means of an electric drill held high overhead.

By 1984, photographic plates disappeared from my life, replaced by electronic CCD detectors and computers. No longer the slow, boring intervals; exposure times were now measured in minutes or even seconds, with computer reductions going on in the background at all times. Even disappointing cloudy nights are used to reduce observations. The pace of observing and of doing astronomy has accelerated manyfold, yet I still opt to observe alone on occasion.

The 30 years I have been observing at Kitt Peak and Cerro Tololo have been marked by a continually advancing instrumental technology. The arts of cutting, baking, and developing plates which I early cultivated so diligently have long since been superseded by electronic and computer processes. But the scientific questions I have attacked have remained remarkably similar. Orbital motions within galaxies, details of nuclear motions, mass distributions within galaxies, and large scale motions of groups of galaxies still remain my areas of research. Rather than use each instrumental advance to search farther into the universe, I have repeatedly used each advance to observe details which were undetectable with earlier instruments.

One thing has not changed: Successful nights at a telescope are among the happiest nights of my life. Most of these nights have been at AURA (Association of Universities for Research in Astronomy) telescopes, and I acknowledge the priceless role that these national facilities have played in my professional career.

A Revised Shapley–Ames Catalog of Bright Galaxies

1982

E very so often in the evolution of a science, a work is produced whose value undoubtedly exceeds all expectations of its authors. Such was *A Survey of the External Galaxies Brighter Than the Thirteenth Magnitude* [1]. (Adelaide Ames, one of its authors and a young researcher at Harvard, died tragically that year in a boating accident.) Since 1932, the Shapley–Ames (SA) catalog, as it is known, has been the principal source of lists of galaxies for study, particularly for systems in the nearby region of the universe. It has been outdated only by the first and second editions of the *Reference Catalog of Bright Galaxies* [2]. Now, Allan Sandage and Gustav Tammann have completely revised and enlarged the Shapley–Ames catalog [3] in an interesting manner.

First, some details of the original work. The catalog is based upon an all-sky survey with a very low plate scale (600 arc seconds per millimeter, taken with 2-inch Ross Tessar and 2-inch Zeiss Tessar lenses). At this scale, the 14-inch-square Palomar Sky Survey plates would be reduced in size to 1.5 inches. At this small plate scale most galaxies appear almost stellar, and their apparent photographic magnitudes can be obtained with remarkable accuracy by comparing the images with reference stars on each plate. The catalogue offered astronomers positions, magnitudes, diameters, and morphological types for a fairly complete set of galaxies with magnitudes down to about 13.2.

Surprisingly, a positional plot of the 1,249 cataloged galaxies showed a decidedly nonrandom arrangement, with about twice as many in the north galactic hemisphere as in the south (Fig. 1). This lopsided distribution ultimately led Gérard de Vaucouleurs to the identification of the Virgo supercluster, a flattened distribution of gravitationally interacting galax-

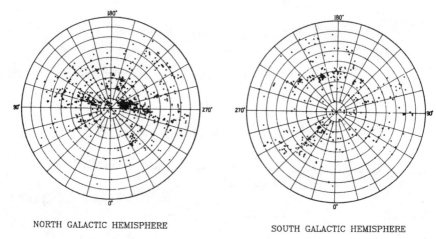

NORTH GALACTIC HEMISPHERE   SOUTH GALACTIC HEMISPHERE

Figure 1. *Distribution of 1025 galaxies brighter than 13th magnitude in the northern (left) and southern (right) galactic hemispheres, from the Shapley–Ames (1932) catalog.*

ies; our galaxy is an outlying member of this system. Just as we observe more stars in the Milky Way in the direction toward the galaxy's center, so we observe more galaxies in the direction toward the center of the super-cluster. This interpretation of the distribution of nearby galaxies must rank as one of the major achievements of twentieth-century astronomy.

There are notable advances in the *Revised Shapley–Ames Catalog* (RSA) in classifications, in velocities, and in completeness [3]. Sandage has re-classified all galaxies, principally from plates of large scale taken with the 200-inch Hale reflector or du Pont 100-inch telescope at Las Campanas, Chile. As a source of consistent morphological types for bright galaxies, the RSA is unsurpassed. For 84 of these galaxies, photographs of superb quality are included. Unlike the large-scale prints and detailed descriptions in the *Hubble Atlas of Galaxies* [4], the RSA mentions the individual galaxy characteristics and the classifying features only briefly. Hence, this volume does not rival the *Hubble Atlas* as a coffee-table conversation piece. However, even at a smaller scale, the photographs are exquisite. For ex-ample, the print of NGC 2146 (the "dusty hand" galaxy of Curtis) is the finest picture of this object I have seen.

Radial velocities for the 1,246 entries are based on over 400 references, all conveniently noted. (Three "galaxies" from the original catalog have been deleted: NGC 643 is a cluster in the Small Magellanic Cloud; NGC 2149 is a galactic nebula, and NGC 6026 a planetary nebula.) These refer-ences are a gold mine containing much more than just velocities; they provide a fine starting point for finding out what is presently known about any galaxy included in the catalog. While the de Vaucouleurs' catalog also had extensive references, it is now six years old, which can be a problem in a rapidly advancing field.

In among all these goodies, what impresses this reader is Appendix A, "Table of Additional Bright Galaxies." Here are 827 galaxies known to be brighter than magnitude 13.4, mostly from the Zwicky survey and hence mostly in the Northern Hemisphere. The inclusion of such a list enlarges the original Shapley–Ames catalog by over 50%, and offers anew a valuable list for observers and statisticians. These are the details by which a science advances.

And what of the galaxies in the RSA? They range from tiny NGC 147 (a flattened dwarf elliptical with an absolute blue magnitude of –14.4) to the supergiant spiral NGC 1961 (a peculiar system resembling the Andromeda galaxy in the relative prominence of its spiral arms and central bulge, with an absolute blue magnitude of –23.7 and a mass 10 times that of the Milky Way), spanning a range of over 5,000 in intrinsic luminosity. In between these two extremes lie the galaxies which populate our region of the universe. For the serious student, amateur or professional, there is no better volume from which to learn about the constituents of our galactic neighborhood.

In an earlier age, we would have been confident that the 2,000 apparently brightest galaxies were telling us about the entire universe. Now it seems wise to recognize that this is but an early step in cataloging only those near to us. When our descendants study galaxies of apparent magnitude 25, they may find it hard not to smile at many of our concepts. Until then, this catalog is a good statement of current knowledge about the brightest galaxies our telescopes can see.

References

1. H. Shapley and A. Ames, *A Survey of the External Galaxies Brighter Than the Thirteenth Magnitude* (Annals Harvard College Observatory, **88**, No. 2, 1932).
2. G. de Vaucouleurs and A. de Vaucouleurs, *Reference Catalogue of Bright Galaxies* (1964), G. de Vaucoulers, A. de Vaucouleurs, and H. G. Corwin, Jr. (1976).
3. A. Sandage and G. A. Tammann, *A Revised Shapley–Ames Catalog* (1984).
4. A. Sandage, *Hubble Atlas of Galaxies* (Carnegie Institution of Washington, 1961).

Star Charts

When Alessandro Piccolomini (1508–1579) wrote his modest book *De le Stelle Fisse Libro Uno* in 1540, he wrote in Italian in a vernacular style, in order to bring his science from ecclesiastical and university confines out to the people. (He also advocated higher education for women.) Today his book is a landmark because it contained the first printed star atlas: 48 woodcut star maps, one for each Ptolemaic constellation. Piccolomini is just one of the nearly 200 entries in *The Sky Explored: Celestial Cartography 1500–1800* [1] by Deborah J. Warner. This encyclopedic work lists alphabetically each author of a star map during this period and gives a few facts concerning his life, details of the publications and of the science contained on the charts, the antecedents of the work, and a bibliography. It does not make for easy reading, but for students of celestial globes, charts, and atlases it is a gold-mine of information and a joy to peruse or to study.

The casual student of maps may know of the two beautiful sky hemispheres (1515) of the Nuremberg artist-mathematician Albrecht Dürer but will now know that the maps were the result of a three-man collaboration of Johann Stabius, who drew the coordinates, Conrad Heinfogel, who positioned the stars, and Dürer, who drew the constellation figures and cut the wood blocks. And the work of Bayer, the Amsterdam Blaeus (father and son), and the London Senex may be familiar. But there is much more to be learned among the unfamiliar. In 1733 Christoph Semler, a Protestant clergyman in Halle, published *Coelum Stellatum*, an atlas of 35 maps showing white stars on black sky. We learn that the copy of Semler's atlas at the Library of Congress has the penciled notation "Given to Whitney Warren ... 1913 and used in designing the ceiling of the New York Central Terminal."

To lovers of old books, collectors, and students of star maps, this is almost as much fun as locating an old star map. It joins two earlier works on the subject, both now collectors' items (but each recently reprinted): E. L. Stevenson's *Terrestrial and Celestial Globes* and Basil Brown's *Astronomical Atlases, Maps and Charts* [2,3]. *The Sky Explored* is expensive, the quality of some of the many reproductions is poor, and, sadly, none are in color. Sources for all reproductions are given, however, and the serious reader could spend many delightful hours searching out the originals in the Library of Congress or in other major collections. But let the reader beware: the search for old globes and star charts is habit forming.

References

1. D. J. Warner, *The Sky Explored: Celestial Cartography 1500–1800* (Liss, New York, 1979).
2. E. L. Stevenson, *Terrestrial and Celestial Globes* (Hispanic Society of America, Yale University Press, New Haven, 1921).
3. B. Brown, *Astronomical Atlases, Maps and Charts* (Search Publishing Co., London, 1937).

Part III
MATTER AND MOTION

M uch extragalactic research during the past two decades has focused on deducing the distribution of matter in the universe and examining the dynamical consequences of this apparently lumpy distribution. Advanced instrumental developments, brilliant theoretical insights, and complex computer simulations together form the underpinnings of these studies. But earlier studies are not irrelevant. Instead, they mark the path that brings us to our discoveries of today. Both the earlier history and some current studies are highlighted in the papers that follow.

"The Local Supercluster and the Anisotropy of the Redshifts" was a talk given at a 1988 Paris celebration for Gérard de Vaucouleurs' 70th birthday. It was published in *The World of Galaxies* (edited by H. G. Corwin, Jr. and L. Bottinelli, Springer-Verlag) in 1989.

"The Rotation of Spiral Galaxies," published in 1983 in *Science*, is a summary of the evidence that orbital velocities of stars about the centers of their galaxies imply the existence of dark matter in and surrounding galaxies.

"Dark Halos Around Spiral Galaxies" was a talk at the First ESO (European Southern Observatory)–CERN Symposium, a meeting to unite astronomers and particle physicists in their discussions of dark matter. It was published in *Large-Scale Structure of the Universe, Cosmology and Fundamental Physics* (edited by G. Setti and L. van Hove, 1984).

"How Much Dark Matter is There?" is a section from a six-chapter book, *Bubbles, Voids, and Bumps in Time* (edited by J. Cornell, Cambridge University Press, 1989). Each of the six chapters has a different author; each author initially presented the material as one of six Lowell Lectures at the Boston Museum of Science and then repeated the lecture a week later at the Smithsonian Institution in Washington DC, 1987.

"A Century of Galaxy Spectroscopy" was delivered as the 1994 Russell Prize Lecture of the American Astronomical Society (AAS) in Tucson, Arizona. It reviews early history and current work relating to complex motions in galaxies, and it appeared in the Astrophysical Journal.

The Local Supercluster and Anisotropy of the Redshifts

1989

I n order to deduce the existence of a Local Supercluster two things are necessary: a knowledge of the structure and a knowledge of the motions within the nearby universe. I find it surprising that the machinery for deducing motions was at hand long before the corresponding machinery for deducing structure, although in some sense both go hand-in-hand, because distance determinations rest very much on measured redshifts. What follows is a very sketchy and very personal understanding of some of the early history.

Early Determination of Motions

Although Hooke suspected as early as 1700 that the sun was moving, the first determination of a stellar proper motion came only in 1718, when Halley [1] established that a few stars had altered their positions since the catalog of Hipparchus over eighteen centuries earlier. In 1779, Lalande predicted that the motion of the sun would not be detectable until several centuries had passed and the sun had moved into a region of different stars. This prediction establishes the "Lalande Principle," a warning to us all to avoid making solemn pronouncements. Only four years later, William Herschel [2] determined the motion of the sun from proper motions of nearby stars. In fairness to Lalande, it is important to note that Herschel used a procedure very different from (and far more clever than) any envisioned by Lalande.

Two features of Herschel's work deserve special mention. First, Herschel understood that the sun could not be at rest: "there is not, in strictness of speaking, one fixed star in the heavens..., when once it is known that some of them are in motion: for the change that must arise by such motions, in the value of a power which acts inversely as the squares of the distances, must be felt in all the neighboring stars; and if these be influenced by the motion of the former, they will again affect those that are next to them, and so on till all are in motion." Second, the Herschel manuscripts in the Royal Astronomical Society archives contain a sketch by Herschel showing the projections onto the equatorial plane of the proper motion components for stars tabulated by Meskelyne and Lalande. From these Herschel deduced the direction of the solar motion. The plot is especially interesting because the positions of the stars are reversed compared to those as seen from the earth [3], suggesting that Herschel had plotted their positions and velocity vectors on a sphere. This procedure is of special interest to me, as I point out below.

The next major step in analyzing extraterrestrial motions came from the incorporation of new principles of physics into astronomical observations. The shift of spectral lines due to a line-of-sight velocity component, discovered by Doppler in 1842 and used by Huggins to obtain the radial velocity of a star in 1868, was applied in 1912 by Slipher to obtain the radial velocity of a galaxy, M31. Since that time, the number of known galaxy redshifts has increased to over 20,000 as increased light-gathering power and more sensitive detectors at visible wavelengths and at 21-cm have speeded up the gathering process. But this early history indicates that by the first part of this century, knowledge existed that would make it possible to map radial velocities of galaxies all across the sky.

Early All-Sky Mapping

Obtaining radial velocities of selected galaxies is an easier task than obtaining an all-sky map of galaxies to a fixed limiting magnitude; understanding the biases and the systematic effects is especially complex for observations made separately from the northern and southern hemispheres. Halley [4] was able to identify six "lucid spots among the fix't stars," of which only the Andromeda galaxy was extragalactic. By 1781, Messier [5] had identified 103 nebulosities; 32 of these were extragalactic, and the great nebula in Andromeda became M31. Most of us know about the telescopic searches of the Herschels, William and son John, which culminated ultimately in the publication of the *New General Catalog* by Dreyer [6]. Both Herschels were aware that the distribution of the spirals was distinct from that of the diffuse nebulosities.

Less well-known now are the studies of the distribution of the nebulae, most of them published in the *Monthly Notices of the Royal Astronomical Society*, which followed the publication of the NGC. Especially attractive are the two color plots of Waters [7]. But almost 50 years were to pass before Shapley and Ames [8] produced a major all-sky catalog which attempted to include only extragalactic objects, to a fixed limiting magnitude. The difference in numbers of bright galaxies in the northern and southern galactic hemispheres (Fig 1, p. 90) was the first major result from this catalog.

In contrast to the 1888–1932 period in which there was little all-sky extragalactic mapping, 1932–1984 was a period of enormous activity. Notable are the Palomar Sky Survey [9] from which Abell [10] extracted a catalog of clusters of galaxies, (as well as its later companions from the Southern Hemisphere); the Shane–Wirtanen [11] counts of northern galaxies to a faint limiting magnitude; the revisions to the Shapley–Ames catalogs by the de Vaucouleurs and their collaborators [12,13]; the catalogs of Zwicky and collaborators [14] and Vorontsov-Velyaminov [15]; the Uppsala catalog of Nilson [16] for northern galaxies larger than 1 arcminute and its southern extension [17]; and the remarkable infrared catalog developed from the IRAS satellite observations [18].

A figure copied from the RC2 (Fig. 1) shows the clustering of galaxies to a plane, evidence cited by de Vaucouleurs for a Local Supercluster. A more recent map has been produced by Lahav [19] based on a combination of the Uppsala and Vorontsov-Velyaminov catalogs. Galaxies larger than 1 arcminute are plotted, and the concentration to a plane extending well across the sky is evident. Even to these greater distances, the sky projection of galaxies is not uniform.

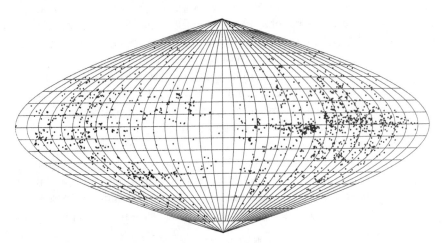

Figure 1. *Distribution of 4364 galaxies from the RC2 [13] in supergalactic coordinates. The central vertical void is due to obscuration from the galactic plane; the north galactic hemisphere is to the right. From Ref. 13, reprinted with permission.*

The Local Supercluster and Velocity Anisotropies

During the decades of the 1950s and 60s, numerous studies, particularly those by de Vaucouleurs, led to general acceptance of a local supercluster, and attention turned next to the anisotropy of velocities induced by the irregular mass distribution. I will trace the history since 1950 by identifying several studies which have played significant roles in defining the structure of our nearby environment, and in examining systematic motions within this region. Though I am not stressing the de Vaucouleurs papers, they make valuable reading as a set, and remind us once again how prescient were many of his ideas.

In 1950, for my Master's thesis, I examined the positions and velocities of the 108 galaxies whose velocities were known, to determine if large scale motions (other than expansion) were present [45]. The analysis involved plotting velocities on a sphere and searching for large regions of excess positive or negative velocities. Like many first steps in science, this early study bears little relation to the sophisticated studies of today, which make use of tens of thousands of galaxies. It did, however, offer de Vaucouleurs a base for his later studies.

The direction which his work was to take can be deduced from an early de Vaucouleurs paper [20]. For a sample of galaxies, he discovered a variation in mean radial velocity along the supergalactic plane for galaxies of increasing apparent magnitude, i.e. distance. The anisotropy in the velocity distribution along the supergalactic plane, evident as the minimum in the Virgo direction in this early work, has been the subject of numerous studies since then, both by de Vaucouleurs and others. Indeed, even today there is no certain agreement as to the magnitude of the mass excess in the direction of the Virgo cluster and how much this excess retards the expansion of the universe at the position of the Local Group.

The detection of the cosmic microwave radiation by Penzias and Wilson [21] not only gave observational support for Big Bang cosmology, it also established a fixed frame of reference (i.e. a sea of photons at 3°K) against which motions could be measured. Initial evidence from Partridge and Wilkinson [22] indicated that the sun had a negligible motion with respect to this frame. But for my colleagues Kent Ford, Norbert Thonnard, Mort Roberts, John Graham, and I, it suggested that the combination of advanced observing techniques and theoretical understanding made the time ripe for a fresh attack on the questions of deviations from a smooth Hubble flow.

For an all-sky sample of over nearly 200 Sc I spirals chosen to be beyond the Virgo cluster, we obtained radial velocities (and magnitudes for the southern objects). The analysis of 96 galaxies with velocities between

3500 and 6500 km s^{-1} [23,24] indicated an anisotropy across the sky, which could be reasonably well fit with a dipole variation. We interpreted this as a motion of the sun of $V_0 = 600 \pm 125$ km s^{-1}, corresponding to a motion of the Local Group of $V_{LG} = 450 \pm 125$ km s^{-1} toward $l = 163° \pm 15°$, $b = -11° \pm 15°$, more than 90° from the direction to the Virgo cluster. This result was generally met with the response that the velocity of the sun could not be this large. Within a year, however, a dipole anisotropy in the microwave background had been detected; the velocity was close to the present value [25,26] for the Local Group of $V_{LG} = 540 \pm 50$ km s^{-1} toward $l = 267° \pm 5°$, $b = +31° \pm 5°$, 44° from the direction to Virgo. Immediately the objection to the results of Rubin et al. changed. Now it was not the magnitude of the optical dipole, but the direction of the apex which was puzzling.

Unfortunately or perhaps fortunately, few other all-sky homogeneous data samples existed in the mid-70's for which distances could be obtained independently of the Hubble velocities. Hence the Rubin et al. sample was subjected to numerous reanalyses during the next ten years [27–30]. Improvements included more sophisticated statistical techniques, new photoelectric magnitudes, better values for foreground extinction, and IRAS infrared galaxy magnitudes. Although it became clear that the original error was underestimated (200 kms^{-1} being a more likely value), the original direction and amplitude remained unaltered. Still, most astronomers were uncomfortable with the result, wished it would disappear, or chose to ignore it.

At about the same time, an examination of the observational consequences of an "infall" (actually a retardation in the expansion) to Virgo was carried out by Peebles [31], followed by creative models and analysis of observational data by Schechter [32] and by Tonry and Davis [33]. These studies, those of Aaronson, Huchra, and Mould [34] plus the work of many others all contributed to produce an atmosphere in which an examination of large scale motions was a legitimate, even trendy, field of research. As Malcolm Longair [35] wrote in his summary of the I. A. U. symposium on *Observational Cosmology* held in Beijing in 1986, "Another hot topic was the observation of large-scale streaming velocities of galaxies. The history of these studies dates back to the much-discussed Rubin and Ford effect and is a good example of an observation, which many people wished would go away, suddenly becoming eminently respectable."

Current Studies of Large-Scale Motions

By the early 1980's, several observational studies were underway in an effort to obtain larger all-sky data sets. A few of these have altered our view of the large-scale structure and motions in the nearby universe: I mention here the CfA plots of the large-scale structure [36,37], the elliptical data and analysis by Faber and colleagues [38], the gravitational field deduced from the IRAS infrared observations [39,40], and the cluster and supercluster studies of Bahcall [41] and her colleagues.

Although evidence had been noted earlier, it was the plots of the Shane–Wirtanen galaxies (Fig. 1, p.8) made by Peebles and colleagues [42], plus the studies of Jôeveer, Einasto, and Tago [43] that shocked astronomers into noticing the large-scale irregularities in the distribution of galaxies. Galaxies were arranged in cells surrounding voids, with stringy filamentary connections between the cells. The discovery of the Boötes void [44] gave additional confirmation to this picture. But it was an initial plot of the distribution of galaxies from the Center for Astrophysics (CfA) redshift survey that showed us how widespread this phenomenon is. The CfA redshift survey is now complete for several more slices, and reemphasizes the spectacular views of voids and filaments. The voids, filaments, and general nonrandom pattern make a striking visual impression. Structures with scales as large as those covered by the observations force us to admit that we have not yet reached a domain where the distribution is smooth.

Given this nonuniform distribution of mass, at least of luminous mass, it is reasonable to expect departures from a smooth Hubble flow. Most recently, the velocity field within about 3000 km s^{-1} has been investigated by a group of seven [38] astronomers, using spectroscopic and photometric data for 400 elliptical galaxies distributed fairly well across the sky. Peculiar velocities relative to the cosmic microwave background indicate a complex pattern of motions: a bulk motion toward a Great Attractor in Centaurus, a small Virgocentric inflow, and a Local Velocity Anomaly.

Conclusions

Our knowledge of the structure and the velocities within the nearby universe continues to grow as observations reveal a greater complexity than had been anticipated. Once again, we are reminded that most of cosmology is observationally led. As befits an observational science, we can observe our progress by examining recent results. I did not expect, during my lifetime, to see a plot of gravitationally induced velocity vectors for galaxies out to distances of 70 Mpc. It may not be a correct map, but it is a magnificently bold first step.

From the most recent studies of distributions and motions in the universe, I propose the following conclusions:

1. Luminous matter in the presently observed universe is distributed in a nonrandom, clumpy manner.

2. This nonuniform distribution of matter gives rise to large-scale bulk motions. Scales of 10's of Mpc and 1000 km s⁻¹ are involved.

3. Lacking evidence to the contrary, we assume that the distributions and motions in the nearby universe can be described by Newtonian gravitational theory.

4. Observations have raised many questions. Even though our answers may not yet be correct, I think we are asking the right questions. Many of these questions were first posed by Gérard de Vaucouleurs, and we acknowledge our scientific and personal debts to him.

Discussion

F. J. Kerr: I was present at the AAS meeting in late 1950 (this was my first AAS meeting). I happened to hear by chance three prestigious council members discussing whether Vera Rubin's paper should be allowed. They seemed to be shocked by this brash young outsider (and female at that) making such an outrageous and impossible suggestion. Fortunately, they did not stop the presentation of the paper [45].

References

1. E. Halley, Phil. Trans. Roy. Soc. London **30**, 737 (1718).
2. W. Herschel, Phil. Trans. Roy. Soc. London **73**, 247 (1783).
3. M. Hoskin, J. Hist. Astron. **11**, 153 (1980).
4. E. Halley, Phil. Trans. Roy. Soc. London **29**, 39 (1715).
5. C. Messier, *Connaissance des temps pour l'année bissextile 1784*, (Paris, 1781), p. 227.
6. J. L. E. Dreyer, Mem. Roy. Astron. Soc. **49**, Part 1 (1888).
7. S. Waters, Mon. Not. Roy. Astron. Soc. **54**, 526 (1894).
8. H. Shapley and A. Ames, Ann. Harvard Coll. Obs. **88** (2) (1932).
9. R. L. Minkowski and G. O. Abell, in *Basic Astronomical Data*, edited by K. Aa. Strand (University of Chicago Press, Chicago, 1963), p. 481.
10. G. O. Abell, Astrophys. J. Suppl. **3**, 211 (1958).
11. C. D. Shane and C. A. Wirtanen, Publ. Lick. Obs. **22**, Part 1 (1967).
12. G. de Vaucouleurs and A. de Vaucouleurs, *Reference Catalog of Bright Galaxies* (RC1) (University of Texas Press, Austin, 1964).
13. G. de Vaucouleurs, A. de Vaucouleurs, and H. G. Corwin, *Second Reference Catalog of Bright Galaxies* (RC2) (University of Texas Press, Austin, 1976).
14. F. Zwicky, E. Herzog, C. T. Kowal, M. Karpowicz, and P. Wild, *Catalog of Galaxies and of Clusters of Galaxies*, **I–VI** (California Institute of Technology, Pasadena, 1961, 1963, 1965, 1966, 1968a, 1968b).
15. B. A. Vorontsov-Velyaminov, V. P. Arkipova, and A. A. Krasnogorska, *Morphological Catalog of Galaxies*, **I–V** (Moscow State University, Moscow, 1962, 1963, 1964, 1968, 1974).

16 P. Nilson, *Uppsala General Catalog of Galaxies*, Uppsala Astron. Obs. Ann. **6**, (1973).

17. A. Lauberts, *The ESO/Uppsala Survey of the ESO (B) Atlas* (European Southern Observatory, Munich, 1982).

18. G. Neugebauer, C. A. Beichman, B. T. Soifer, H. H. Aumann, T. J. Chester, T. N. Gautier, F. C. Gillet, M. G. Hauser, J. R. Houck, C. J. Lonsdale, F. J. Low, and E. T. Young, Science **224**, 14 (1984).

19. D. Lynden-Bell and O. Lahav, in *Large-Scale Motions in the Universe: A Vatican Study Week*, edited by V. C. Rubin and G. V. Coyne (Princeton University Press, Princeton, 1988), p. 199.

20. G. de Vaucouleurs, Astron. J. **63**, 253 (1958).

21. A. A. Penzias and R. W. Wilson, Astrophys. J. **142**, 419 (1965).

22. R. B. Partridge and D. T. Wilkinson, Phys. Rev. Lett. **18**, 557 (1967).

23. V. C. Rubin, W. K. Ford, Jr., N. Thonnard, and M. Roberts, Astron. J. **81**, 719 (1976).

24. V. C. Rubin, W. K. Ford, Jr., N. Thonnard, M. Roberts, and J. A. Graham, Astron. J. **81**, 687 (1976).

25. D. J. Fixsen, E. S. Cheng, and D. T. Wilkinson, Phys. Rev. Lett. **50**, 620 (1983).

26. P. M. Lubin, G. Villela, G. L. Epstein, and G. F. Smoot, Astrophys. J. Lett. **298**, L1 (1985).

27. P. L. Schechter, Astron. J. **82**, 569 (1977).

28. P. Jackson, unpublished Ph. D. thesis (University of California, Santa Cruz, 1982).

29. C. J. Peterson and W. Baumgart, Astron. J. **91**, 530 (1986).

30. C. A. Collins, R. D. Joseph, and N. A. Robertson, in *Galaxy Distances and Deviations from Universal Expansion*, edited by B. F. Madore and R. B. Tully (Reidel, Dordrecht, 1986), p. 131.

31. P. J. E. Peebles, Astrophys. J. **205**, 318 (1976).

32. P. L. Schechter, Astron. J. **85**, 801 (1980).

33. J. L. Tonry and M. Davis, Astrophys. J. **246**, 680 (1981).

34. M. Aaronson, J. Huchra, and J. Mould, Astrophys. J. **229**, 1 (1979).

35. M. Longair, in *Observational Cosmology*, edited by A. Hewitt, G. Burbidge, and L. Z. Fang (Reidel, Dordrecht, 1987), p. 823.

36. V. de Lapparent, M. J. Geller, and J. P. Huchra, Astrophys. J. Lett. **202**, L1 (1986).

37. M. Geller and J. P. Huchra, in *Large-Scale Motions in the Universe: A Vatican Study Week*, edited by V. C. Rubin and G. V. Coyne (Princeton University Press, Princeton, 1988), p. 3.

38. S. M. Faber and D. Burstein, in *Large-Scale Motions in the Universe: A Vatican Study Week*, edited by V. C. Rubin and G. V. Coyne (Princeton University Press, Princeton, 1988), p. 115.

39. A. Yahil, in *Large-Scale Motions in the Universe: A Vatican Study Week*, edited by V. C. Rubin and G. V. Coyne (Princeton University Press, Princeton, 1988), p. 219.

40. M. A. Strauss and M. Davis, in *Large-Scale Motions in the Universe: A Vatican Study Week*, edited by V. C. Rubin and G. V. Coyne (Princeton University Press, Princeton, 1988), p. 255.

41. N. Bahcall, Ann. Rev. Astron. Astrophys. **26**, 631 (1988).

42. M. Seldner, B. Siebers, E. J. Groth, and P. J. E. Peebles, Astron. J. **82**, 249 (1977).

43. M. Jôeveer, J. Einasto, and E. Tago, Mon. Not. Roy. Astron. Soc. **185**, 357 (1978).

44. R. P. Kirshner, A. Oemler, Jr., P. L. Schechter, and S. A. Shectman, Astrophys. J. Lett. **248**, L57 (1981).

45. V. C. Rubin, Astron. J. **56**, 47 (1951).

The Rotation of Spiral Galaxies

1983

Historians of astronomy may someday call the mid-twentieth century the era of galaxies. During those years, astronomers made significant progress in understanding the structure, formation, and evolution of galaxies. I will discuss a few selected early steps in determining the internal dynamics of galaxies, as well as later research concerning the rotation and mass distribution within galaxies. These studies have contributed to the present view that much of the mass of the universe is dark.

In 1845, Lord Rosse [1] constructed a mammoth 72-inch reflecting telescope, which remained the largest telescope in the world until the construction of the "modern" 100-inch telescope on Mount Wilson in 1917. Unfortunately, the Rosse telescope did not automatically compensate for the rotation of the earth, so an object set in the field of the telescope would race across the field. Nevertheless, Lord Rosse made the major discovery that some nebulous objects show a spiral structure [2].

Insight into the nature of these enigmatic objects was slow in coming, but by 1899 the combination of good tracking telescopes and photographic recording made it possible for Scheiner [3] to obtain at the Potsdam Observatory a spectrum of the nucleus of the giant spiral in Andromeda, M31. From the prominence of the H and K absorption lines of calcium, Scheiner correctly concluded that the nucleus of M31 was a stellar system composed of stars like the sun, rather than a gaseous object.

Even earlier, Lord and Lady Huggins [4] had attempted at their private observatory to obtain a spectrum of M31, but their labors were fraught with disaster. Exposure times were so long that each fall when M31 was overhead they would expose consecutively on one plate during the early evening hours, until they could build up the requisite exposure (perhaps 100 hours). They would then superpose a solar spectrum adjacent to the galaxy spectrum to calibrate the wavelength scale. Year after year, their single exposure was a failure, sometimes showing no galaxy spectrum, sometimes contaminated by the solar spectrum; once the plate was accidentally placed emulsion side down, where it dried stuck to the laboratory table. A copy of their 1888 failure is published along with their remarkable stellar spectra [4].

Slipher [5], working with the 24-inch reflector at Lowell Observatory, initiated in 1912 systematic spectral observations of the bright inner regions of the nearest galaxies. For NGC 4594, the Sombrero galaxy, he detected inclined lines [6; see also 7], which he correctly attributed to stars collectively orbiting about the center of that galaxy. On the side of the nucleus where the rotation of the galaxy carries the stars toward the observer, spectral lines are shifted toward the blue region of the spectrum with respect to the central velocity. On the opposite side, where rotation carries the stars away from the observer, lines are shifted toward the red spectral region.

Several years later, Pease [8] convincingly illustrated that rotation *was* responsible for the inclined lines. Using the 60-inch telescope on Mount Wilson during the months of August, September, and October 1917, he patiently acquired a 79-hour exposure of the spectrum of the nuclear region of M31 with the slit aligned along the apparent major axis of the galaxy. Inclined lines appeared. A second exposure made during these 3 months in 1918, with the slit placed perpendicular to the major axis along the apparent minor axis, showed no inclination of the lines. Along the minor axis, stars move at right angles to the line-of-sight of the observer, so no Doppler shift appears. The rotation of galaxies was established even before astronomers understood what a galaxy was.

By the mid-1920's, astronomers knew that we lived in a galaxy composed of billions of stars distributed principally in a disk, all rotating about a distant center. Each spiral viewed in a telescope, located at an enormous distance beyond our galaxy, is itself a gigantic gravitationally bound system of billions of stars all orbiting in concert about a common center. This knowledge came from a variety of observational discoveries: Henrietta Leavitt's [9] discovery of Cepheid variables, stars whose periods of light fluctuation reveal their true brightnesses and hence distances; Shapley's [10] demonstration that the globular clusters surround our own galaxy like a halo, with a geometric center located not at the sun but at a great distance in the dense star clouds of Sagittarius; and Hubble's [11] discovery of Cepheids in M31 and M33.

In a brilliant study, Öpik [12] used the rotational velocities in M31 to estimate its distance. His result, 450,000 parsecs (1 pc = 3 × 10^{13} km), comes closer to the distance in use today, 750,000 pc, than the distance of

230,000 pc which Hubble derived from the Cepheids and which he was still using in 1950 [13]. Öpik's distance was based on the procedure used currently to determine the masses of galaxies. For a particle of mass m moving in a circular orbit with velocity V about a spherical distribution of mass M at distance r from the center, the equality of gravitational and centrifugal forces gives

$$\frac{GM(r)m}{r^2} = \frac{mV^2(r)}{r} \tag{1}$$

where G is the constant of gravitation and $M(r)$ is the total mass contained out to a distance r from the center. It follows that the mass interior to r is given by

$$M(r) = rV^2(r) / G \tag{2}$$

or

$$V(r) = [GM(r) / r]^{1/2}. \tag{3}$$

Today we adopt a distance to a galaxy (generally from its Hubble velocity), determine the rotational velocity V at each r, and then calculate the variation of M with r. Öpik assumed that the ratio of mass to luminosity was the same for M31 as for our galaxy, and used M and V to determine r. Both the dynamics and the astrophysics were sound, and the result convinced many astronomers that spiral galaxies were external to our own Milky Way system.

Modern Optical Observations of Spiral Galaxy Rotation

Several observational procedures are available today to study the rotation (that is, orbiting) of stars and gas in a spiral galaxy. A spectrograph with a long slit will record the light arising from all of the stars in the galaxy along each line-of-sight. The resulting spectrum will be composite, the sum of individual stellar spectra. A stellar motion toward the observer will displace each spectral line toward the blue region of the spectrum; a motion away from the observer will displace each line toward the red. Measurement of the successive displacements along a spectral line will give the mean velocity of the stars corresponding to that location in the galaxy. The shape and width of the line contain information concerning the random motions along the line-of-sight.

Starlight from external galaxies is generally too faint to permit a dense spectral exposure on which accurate positional measurements may be made from stellar absorption lines. One way around this difficulty is to measure velocities from the emission lines arising in the ionized gas clouds surrounding the hot young blue stars which delineate the spiral structure. Because the light from these clouds is emitted principally in a few lines of abundant elements (hydrogen, ionized oxygen, ionized nitrogen, and ionized sulfur), a measurable exposure is obtained in a fraction of the time required for the stellar exposure. This is the method we employ in our optical observations. A third procedure is to observe in the radio spectral region, at the wavelength of 21 cm emitted by the hydrogen atom. This radiation is not obscured by the prominent dust lanes seen in spiral galaxies and becomes the object of observation for the radio astronomer.

Following the pioneering work discussed above, observations of the rotation of galaxies proceeded very slowly. A major addition to our knowledge came from the extensive work of Margaret and Geoffrey Burbidge [14]. Due to the long observing times, velocities were obtained only for the brightest inner regions. Only for a few of the nearest galaxies was it possible to observe individual regions and map the rotation to large nuclear distance [15]. Until recently, radio observations had limited spatial resolution and hence poor velocity accuracy. Astronomers attempting to examine the dynamical properties of galaxies [16] had few measured velocities well beyond the bright nuclear regions.

For the past several years, W. K. Ford, Jr., N. Thonnard, D. Burstein, B. Whitmore, and I [17–19] have been using modern detectors to study the dynamical properties of isolated spiral galaxies. We attempt to measure the rotational velocities across the entire optical galaxy. We have observed galaxies of Hubble types Sa (disk galaxies with large central bulges, and tightly wound weak arms with few emission regions), Sb, and Sc (disk galaxies with small central bulges, and open arms with prominent emission regions), emphasizing galaxies of high and low luminosity within each class. Our aim is to learn how rotational properties vary along the Hubble sequence of galaxies, and within a Hubble class, and to relate the dynamical properties to other galaxy parameters.

For 60 Sa, Sb, and Sc program galaxies, we have obtained spectra with the 4-meter telescopes at Kitt Peak National Observatory (near Tucson, Arizona) and Cerro Tololo Inter-American Observatory (near La Serena, Chile). A few spectra come from the 2.5-meter du Pont Telescope of the Carnegie Institution of Washington at Las Campanas (also near La Serena). Optical observations are made with spectrographs which incorporate an RCA C33063 image tube [20]. The image is photographed from the final phosphor of this tube. Use of this electronic enhancement device makes it possible to obtain spectra at a high spatial scale and of high velocity accu-

racy with exposure times of about three hours, on Kodak IIIa-J plates which
have been baked in forming gas and preflashed to enhance their speed.

We choose galaxies that are relatively isolated, that are not strongly
barred, that subtend an angular size at the telescope which approximately
matches the length of the spectrograph slit, and that are viewed at relatively high inclination to minimize uncertainties in transforming line-of-
sight velocities to orbital velocities in the galaxy. Distances are calculated
from velocities arising from the cosmological expansion, adopting a Hubble
constant of 50 km sec^{-1} Mpc^{-1}.

A sample of Sc program galaxies and spectra is shown in Fig. 1; the
spectra were taken with the spectrograph slit aligned along the galaxy major
axis. The galaxies are arranged by increasing luminosity. Velocities are
determined by measuring the displacement of the emission lines with respect to the night sky lines to an accuracy of 1 μm. We measure with a
microscope which moves in two dimensions, a rather old-fashioned but
still very accurate technique. Many astronomers now trace plates with a
microdensitometer or obtain digital data directly at the telescope. From
the measured velocities on each side of the nucleus a mean curve is formed.
It describes the circular velocity in the plane of the galaxy as a function of
distance from the nucleus, as shown on the right in Fig. 1.

Within a Hubble type, rotation curves vary systematically with luminosity. For a low-luminosity galaxy (NGC 2742), velocities rise gradually
from the nucleus and reach a low maximum velocity only in the outer
regions. For a high-luminosity galaxy (UGC 2885), rotational velocities
rise steeply from the nucleus and reach a "nearly flat" portion in a small
fraction of the galaxy radius. An Sc of low luminosity (6×10^9 solar luminosities) has a maximum rotational velocity V_{max} near 100 km sec^{-1}, compared with V_{max} near 225 km sec^{-1} for an Sc of high luminosity (2×10^{11}
solar luminosities). Dynamical variations from one Hubble class to another are equally systematic. What differentiates Sa and Sc galaxies of
equal luminosity are the amplitudes of their rotational velocities. At equal
luminosity, an Sa has a higher rotational velocity than an Sb, which in turn
has a higher velocity than that of an Sc. A low luminosity Sa has $V_{max} = 175$
km sec^{-1}, a high-luminosity Sa has $V_{max} = 375$ km sec^{-1}. Each value is
higher than that of the corresponding Sc galaxy.

Mass Distribution Within Spiral Galaxies

What do these flat rotation curves tell us about the mass distribution
within a disk galaxy? Unfortunately, we cannot determine a unique mass
distribution from an observed rotation curve; only the mass distribution

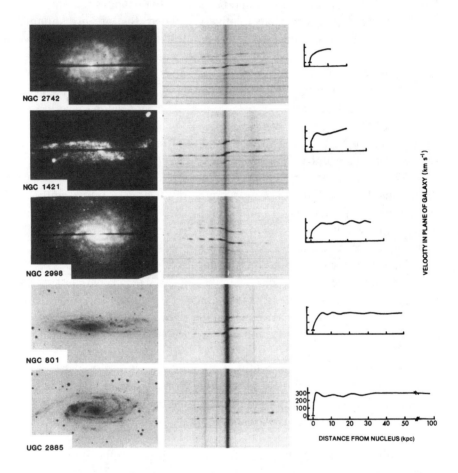

Figure 1. Spectra and rotation curves for five Sc galaxies, arranged according to increasing luminosity. Photographs for NGC 2742, 1421, and 2998 are copies of the television screen which displays the image reflected off the spectrograph slit jaws. The dark line crossing the galaxy is the spectrograph slit. NGC 801 and UGC 2885 are reproduced from plates taken at the prime focus of the 4-meter telescope at Kitt Peak National Observatory by B. Carney. The corresponding spectra are arranged with wavelength increasing from the bottom to the top. The strongest step-shaped line in each spectrum is from hydrogen in the galaxy, and is flanked by weaker lines of forbidden ionized nitrogen. The strong vertical line in each spectrum is the continuum emission from stars in the nucleus. The undistorted horizontal lines are emission from the earth's atmosphere, principally OH. The curves at the right show the rotational velocities as a function of nuclear distance, measured from the emission lines in the spectra.

for an assumed model can be deduced. In practice, a spheroidal or disk model is adopted, and an integral equation is solved for the density distribution as a function of r [21].

In several cases of special interest, the velocity or mass distribution follows directly from Eq. (2) or Eq. (3), for a simple spheroidal model:

1. *Central mass.* In the solar system, where essentially all of the mass is in the sun, $M(r)$ is constant for all distances beyond the sun. Velocities of all planets decrease as $r^{-1/2}$, that is, a Keplerian decrease, with increasing distance from the sun [Eq. (3)]. Mercury (solar distance = 0.39 of the earth's distance) orbits with a velocity of 47.9 km sec^{-1}. Pluto, 100 times farther (39 earth distances), moves with a velocity one-tenth that of Mercury, $V = 4.7$ km sec^{-1}. For galaxies, the lack of a Keplerian falloff in velocities indicates that most of the mass is not located in the nuclear regions.

2. *Solid body.* For a body of uniform density, $M(r) \propto r^3$. Thus $V(r)$ increases linearly with r; the object will exhibit a constant angular velocity which produces solid body rotation. Some galaxies show linear velocity curves near the nucleus, although the spatial resolution is generally poor. Solid body rotation is rare in the outer regions of disk galaxies.

3. *Constant velocity.* For $V(r)$ constant, $M(r)$ increases with r. This variation describes the velocities we observe in galaxies. Density, $M(r)/r^3$, decreases with increasing r as $1/r^2$ for spherical models. Although the density is decreasing outward, the mass in every concentric shell r is the same for all r. Hence a rotation curve which is flat (or slightly rising) out to the final measurable point indicates that the mass is not converging to a limiting mass at the present observational limit of the optical galaxy for a galaxy modeled as a single spheroid.

The mass interior to r is calculated from the observed velocity and Eq. (2). Masses range up to several times 10^{12} solar masses for the most luminous spirals. Because rotational velocity (at a fixed luminosity) is higher in an Sa than in an Sb, and higher in an Sb than in an Sc, big-bulge spirals (Sa's) have a higher mass density (V^2/r^2) at every r than Sb's and Sc's of the same luminosity. Moreover, the distribution of mass with radius is similar, within a single scale factor, for all of the Sc's, most of the Sb's, and a few of the Sa's.

In the disk of a normal spiral galaxy, surface brightness is observed to decrease exponentially with increasing radial distance, while the flat rotation curves imply that mass density falls slower, as $1/r^2$. Hence locally the ratio of mass to luminosity M/L increases with increasing distance from the nucleus. If we average the mass and the luminosity across the entire visible galaxy, then the ratio of mass to luminosity is a function of Hubble type, but is independent of luminosity within a type. This statement holds for observed ranges of (blue) luminosity L_B up to a factor of 100. Consequently, the average mass per unit (blue) luminosity in a galaxy, a measure of the stellar population in the galaxy, is a good indicator of Hubble type.

These observational results are summarized in Fig. 2 and as follows. Within a Hubble type, high-luminosity galaxies are larger, have higher mass and density, have stellar orbital velocities which are larger, but have the same value of M/L_B within the isophotal radius as do low-luminosity spirals. At equal luminosity, Sa, Sb, and Sc galaxies are the same size, but the Sa has a higher density, larger mass, higher rotational velocity, and larger value of M/L_B. This interplay of luminosity and Hubble type has convinced us that galaxies cannot be described by a single parameter sequence, be it Hubble type or luminosity or mass. At least two parameters are necessary; Hubble type and luminosity are one such pair.

Nonluminous Mass in Galaxies

Observations of rotation curves which do not fall support the inference that massive nonluminous halos surround spiral galaxies. The gravitational

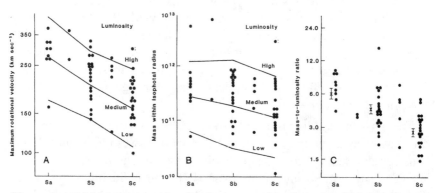

Figure 2. (A) Maximum rotational velocity as a function of Hubble type for the galaxies studied. Lines show the increase in V_{max} with earlier Hubble type for galaxies of low, intermediate, and high luminosity. (B) Mass (solar masses) within the optical galaxy as a function of Hubble type. Within a Hubble type, higher velocity and higher mass indicate higher luminosity. (C) Mass-to-blue luminosity ratio (solar units) as a function of Hubble type. The mean value and the 1σ range are indicated. Within a Hubble type, there is no correlation of the mass-to-luminosity ratio with luminosity.

attraction of this unseen mass, much of it located beyond the optical image, keeps the rotational velocities from falling. The increase in M/L_B with nuclear distance indicates that this nonluminous mass is much less concentrated toward the center of the galaxy than are the visible stars and gas. Even at large nuclear distances, the average density of the nonluminous matter is several orders of magnitude greater than the mean density of mass in the universe. Hence it is clumped around galaxies and is not just an extension of the overall background density.

During the past decade, there has been growing acceptance of the idea that perhaps 90% of the mass in the universe is nonluminous [22]. The requirements for such mass come from a variety of observations on a variety of distance scales. In our own galaxy, we can calculate the disk mass at the position of the sun by observing the attraction of the disk on stars high out of the plane, and comparing this mass with that which we can enumerate in stars and gas. Oort [23] first showed that the counted density at the sun is less by about one-third than that implied by the dynamics, 0.15 solar mass per parsec cubed. But self-gravitating disks of stars are unstable against formation of barlike distortions. Hence it is appealing to place the unseen matter in a halo, rather than a disk, and solve both the problem of the Oort limit and that of the stability of spiral disks. Moreover, a halo of low density at the solar position is consistent both with the Oort limit and with a halo which becomes dominant beyond the optical galaxy.

Additional evidence for a massive halo surrounding our galaxy comes from individual stars and gas at greater nuclear distance than the sun, whose rotational velocities continue to rise [24]. The galaxy rotational velocity is high, befitting the enormous distances. The sun, carrying the planets with it, orbits the galaxy with a speed of 220 km sec^{-1} (500,000 miles per hour); even so, it takes us 225 million years to complete one revolution.

On scale lengths of the order of galaxies, $20 < R < 100$ kpc, the flat rotation curves observed at both optical [17,19,25] and radio [26] wavelengths and the consequent increase with radius of the mass-to-luminosity ratio offer supporting evidence for heavy halos. Moreover, the dynamics of the globular clusters [27] and of the dwarf satellite galaxies [28] which orbit our galaxy imply a mass which increases approximately linearly with increasing radius, to distances as great as 75 or 100 kpc.

On distance scales as great as hundreds to thousands of kiloparsecs, there is equally impressive evidence that the gravitational mass far exceeds the luminous mass. The evidence comes from the orbits of binary galaxies [29] (our galaxy and the Andromeda galaxy form one such pair), random motions of galaxies in clusters [30], and the distribution of hot gas in clusters of galaxies [31] as observed by the x-ray emission. Such observations compose a body of evidence which lends support to the concept of heavy halos clumped around individual galaxies.

Increased interest in rotation curves has led astronomers to related studies. Radio astronomers have determined rotational velocities beyond the optical image for a few galaxies in which the neutral hydrogen distribution extends farther than the optical galaxy. Sancisi [32] suggests that rotation curves do fall beyond the optical galaxy, perhaps by 10% in velocity over a small radial distance, but then level off once again. Casertano [33], Bahcall and collaborators [34], and Caldwell and Ostriker [35] have constructed multicomponent mass distributions, consisting of central point masses, disks (sometimes truncated), bulges, and halos, which can reproduce the observed rotational properties of spiral galaxies.

Observations in a variety of spectral ranges have been unable to detect the dark matter. Massive halos do not appear to radiate significantly in the ultraviolet, visible, infrared, or x-ray regions; they are not composed of gas or of normal low-luminosity stars. Other possible forms include fragments of matter which never became luminous (Jupiter-like planets, black holes, neutrinos, gravitinos, or monopoles) [36]. The enormity of our ignorance can be measured by noting that there is a range in mass of 10^{70} between non-zero-mass neutrinos and massive black holes. Not until we learn the characteristics and the spatial distribution of the dark matter can we predict whether the universe is of high density, so that the expansion will ultimately be halted and the universe will start to contract, or of low density, so that the expansion will go on forever.

Theoretical models of the formation and evolution of galaxies currently attempt to take into account the environmental effects of enormous quantities of nonluminous matter surrounding newborn and evolving galaxies. Most likely, the formation processes for galaxies and for the nonluminous matter were distinct; one attractive idea is that much of the mass of the universe was arranged in its nongaseous dark guise before the galaxies began to form. Recent reviews by Rees [36], Silk [37], and White [38] enumerate the various interesting possibilities. An especially pleasing model by Gunn [39] suggests that disks of spirals form from relatively slowly infalling material. Regardless of which evolutionary scheme ultimately explains the current phases of galaxies and of the universe, we have learned that galaxies are not the isolated "island universes" imagined by Hubble [40], but that environmental effects play a prominent role in directing their evolution.

For nonbelievers in heavy halos and invisible mass, an alternative explanation [41] modifies the $1/r^2$ dependence [Eq. (1)] in Newton's law of gravitation for large r. Under this circumstance, the distribution of mass follows the distribution of light, but the velocities resulting from the mass distribution remain high due to a modified law of gravitation. For the present, this possibility must remain as a last resort.

Why Was It Thought that Rotational Velocities Decreased?

Most present-day astronomers grew up believing that disk galaxies had Keplerian velocities at moderate distances. The reasons for this belief are easy to identify. In the early 1900's, astronomers were more at home with the planets than with the galaxies. Slipher's early observations concerned the study of the planets, Percival Lowell's major interest. Galaxies were studied only to learn whether these nebulous disks in the sky were the stuff from which planets were formed. Slipher used Saturn as a radial velocity standard for his M31 spectra; he characterized the spectrum of the Sombrero galaxy as "planetary." Astronomers were predisposed to draw an analogy between the distribution of luminosity in a galaxy and the distribution of mass in the solar system.

Early spectral observations of galaxies generally did not extend beyond the bright nuclear bulges. Measured velocities usually increased with increasing nuclear distances and mass distributions were in the solid body domain. By 1950, only a handful of rotation curves had been determined, and astronomers were acting on their expectations when they "saw" a region of Keplerian falling velocities. Mayall [42] described his newly determined velocities in M33 as follows: "As in the Andromeda nebula, the results suggested more or less constant angular velocity ... for the main body of the spiral. Beyond the main body these outer parts rotate more slowly ... as in a planetary system where most of the mass lies inside the orbit concerned." De Vaucouleurs [43] reviewed the eight available rotation curves in 1959 and concluded, "In all cases the rotation curve consists of a straight inner part in the region of constant angular velocity up to a maximum at $R = R_m$ beyond which the rotational velocity decreases with increasing distance to the center and tends asymptotically toward Kepler's third law." But with the 20/20 vision of hindsight, plots of the data reveal only a scatter of points, from which no certain conclusion can be drawn.

Yet not all astronomers were blind to the contradiction posed by rapidly falling light distributions and nonfalling rotational velocities. In a classic paper discussing the structure and dynamics of NGC 3115, an almost featureless disk S0 galaxy, the ever-wise Oort [44] wrote, "It may be concluded that the distribution of mass in the system must be considerably different from the distribution of light." And he concluded, "The strongly condensed luminous system appears imbedded in a large and more or less homogeneous mass of great density." Schwarzschild [45], too, looked at the rotation curve of M31 and noted that "Contrary to earlier indication, the five normal points in Fig. 1 do not suggest solid body rotation.... Rather, they suggest fairly constant circular velocity over the whole interval from 25 arcminutes to 115 arcminutes (5–25 kpc with the current distance)."

Indeed, the literature is replete with isolated comments stressing the same point. When in 1962 Rubin *et al.* [46] examined the kinematics of 888 early type stars in our Galaxy, they concluded that beyond the sun "the rotation curve is approximately flat. The decrease in rotational velocity expected for Keplerian orbits is not found." Shostak [47] noted that "the overwhelming characteristic of the velocity field of NGC 2403 is the practically constant circular rotation seen over much of the object."

Now that systematic studies indicate that virtually all spiral galaxies have rotational velocities which remain high to the limits of the optical galaxy, we can recognize the circuitous route which brought us to this knowledge. Astronomers can approach their tasks with some amusement, recognizing that they study only the 5% or 10% of the universe which is luminous. Future astronomers will have to be clever in devising detectors which can map and study this ubiquitous matter which does not reveal itself to us by its light.

References and Notes

1. Third Earl of Rosse, Br. Assoc. Adv. Sci. Rep. (1884), p. 79; P. Moore, *The Astronomy of Birr Castle* (Mitchell Beazley, London, 1971).
2. Third Earl of Rosse, Philos. Trans. R. Soc. London (1850), p. 110.
3. J. Scheiner, Astrophys. J. **9**, 149 (1899).
4. Sir William Huggins and Lady Huggins, *An Atlas of Representative Spectra* (Wm. Wesley and Son, London, 1899), plate **II**.
5. V. M. Slipher, Lowell Obs. Bull. No. 58 (1914).
6. V. M. Slipher, *ibid.* No. 62 (1914).
7. M. Wolf, Vierteljahresschr. Astron. Ges. **49**, 162 (1914).
8. F. G. Pease, Proc. Natl. Acad. Sci. U.S.A. **4**, 21 (1918).
9. H. Leavitt, Harvard Coll. Obs. Ann. **60**, 87 (1908); Harvard Coll. Obs. Circ. No. 179 (1912).
10. H. Shapley, Proc. Astron. Soc. Pacific **30**, 42 (1918); Astrophys. J. **48**, 154 (1918).
11. E. Hubble, Observatory **43**, 139 (1925).
12. E. Öpik, Astrophys. J. **55**, 406 (1922).
13. E. Hubble, *The Realm of the Nebulae* (Yale Univ. Press, New Haven, Conn., 1936), p. 134; E. Holmberg, Medd. Lunds Astron. Obs. Ser. 2 (No. 128) (1950).
14. E. M. Burbidge and G. R. Burbidge, in *Galaxies and the Universe*, edited by A. Sandage, M. Sandage, and J. Kristian (Univ. of Chicago Press, Chicago, 1975), p. 81.
15. V. C. Rubin and W. K. Ford, Jr., Astrophys. J. **159**, 379 (1970).
16. P. Brosche, Astron. Astrophys. **23**, 259 (1973).

17. V. C. Rubin, W. K. Ford, Jr., and N. Thonnard, Astrophys. J. Lett. **225**, L107 (1978); V. C. Rubin and W. K. Ford, Jr., Astrophys. J. **238**, 471 (1980).

18. D. Burstein, V. C. Rubin, N. Thonnard, and W. K. Ford, Jr., Astrophys. J. **253**, 70 (1982).

19. V. C. Rubin, W. K. Ford, Jr., and N. Thonnard, *ibid.* **261**, 439 (1982).

20. This tube is known throughout the astronomical community as the "Carnegie" image tube, for its development was the product of a cooperative effort by RCA and a National Image Tube Committee, under the leadership of M. Tuve and W. K. Ford, Jr., of the Carnegie Institution of Washington, funded by the National Science Foundation.

21. L. Perek, Adv. Astron. Astrophys. **1**, 165 (1962); A. Toomre, Astrophys. J. **138**, 385 (1963). See also Burbidge and Burbidge [14].

22. J. P. Ostriker, P. J. E. Peebles, and A. Yahil, Astrophys. J. Lett. **193**, L1 (1974); J. Einasto, A. Kaasik, and E. Saar, Nature (London) **250**, 309 (1974). For a dissenting view, see G. R. Burbidge, Astrophys. J. Lett. **196**, L7 (1975).

23. J. Oort, Bull. Astron. Inst. Neth. **15**, 45 (1960); in *Galaxies and the Universe*, edited by A. Sandage, M. Sandage, and J. Kristian (Univ. of Chicago Press, Chicago, 1975), p. 455.

24. S. Kulkarni, L. Blitz, and C. Heiles, Astrophys. J. Lett. **259**, L63 (1982).

25. S. M. Faber and J. S. Gallagher, Annu. Rev. Astron. Astrophys. **17**, 135 (1979).

26. M. S. Roberts and A. H. Rots, Astron. Astrophys. **26**, 483 (1973); A. Bosma, thesis, Rijksuniversiteit te Groningen (1978); Astron. J. **86**, 1791 (1981); *ibid.*, p. 1825.

27. F. D. A. Hartwick and W. L. W. Sargent, Astrophys. J. **221**, 512 (1978).

28. J. Einasto, in [22]; D. Lynden-Bell, in *Astrophysical Cosmology*, edited by H. A. Bruck, G. V. Coyne, and M. S. Longair (Pontifical Academy, Vatican City State, 1982), p. 85.

29. S. D. Peterson, Astrophys. J. **232**, 20 (1979).

30. F. Zwicky, Helv. Phys. Acta **6**, 110 (1933); H. J. Rood, Astrophys. J. Suppl. **49**, 111 (1982); W. Press and M. Davis, Astrophys. J., in press.

31. D. Fabricant, M. Lecar, and P. Gorenstein, Astrophys. J. **241**, 552 (1980); W. Forman and C. Jones, Annu. Rev. Astron. Astrophys. **20**, 547 (1982).

32. R. Sancisi, private communication; R. Sancisi and R.J. Allen, Astron. Astrophys. **74**, 73 (1979).

33. S. Casertano, Mon. Not. R. Astron. Soc., in press.

34. J. N. Bahcall, M. Schmidt, and R. M. Soneira, Astrophys. J. Lett. **258**, L23 (1982).

35. J. A. R. Caldwell and J. P. Ostriker, Astrophys. J. **251**, 61 (1982).

36. M. J. Rees, in *Observational Cosmology* (Geneva Observatory, Geneva, Switzerland, 1978), p. 259; in *Astrophysical Cosmology*, edited by H. A. Bruck, G. V. Coyne, and M. S. Longair (Pontifical Academy, Vatican City State, 1982), p. 3.

37. J. Silk, in *Astrophysical Cosmology*, edited by H. A. Bruck, G. V. Coyne, and M. S. Longair (Pontifical Academy, Vatican City State, 1982), p. 427; see also S. M. Faber, in *ibid.*, p. 191; J. P. Ostriker, in *ibid.*, p. 473.

38. S. White, in *Morphology and Dynamics of Galaxies* (Geneva Observatory, Geneva, Switzerland, 1982), p. 291.

39. J. E. Gunn, in *Astrophysical Cosmology*, edited by H. A. Bruck, G. V. Coyne, and M. S. Longair (Pontifical Academy, Vatican City State, 1982), p. 233.

40. E. Hubble, in *The Realm of the Nebulae* (Yale Univ. Press, New Haven, Conn., 1936), p. 98.

41. J. Bekenstein, Int. Astron. Union Symp., in press; J. Bekenstein and M. Milgrom, preprint.

42. N. U. Mayall [Publ. Obs. Univ. Mich. No. 10 (1950), p. 9] notes that the form of the rotation curve of M31, compared with that of our Galaxy, suggests that the distance of M31 (then adopted as 230,000 pc) is too small. The use of rotational properties to establish distances is only now coming into wide use.

43. G. de Vaucouleurs, in *Handbuch der Physik*, edited by S. Flugge (Springer, Berlin, 1959), vol. 20, p. 311.

44. J. H. Oort, Astrophys. J. **91**, 273 (1940).

45. M. Schwarzschild, Astron. J. **59**, 273 (1954).

46. V. C. Rubin, J. Burley, A. Kiasatpoor, B. Klock, G. Pease, E. Rutscheidt, and C. Smith, *ibid.* **67**, 527 (1962).

47. G. S. Shostak, Astron. Astrophys. **24**, 411 (1973).

48. More complete references are contained in the literature cited. I thank my colleagues Kent Ford and Norbert Thonnard for their continued assistance and support in all the phases of this work and Drs. Robert J. Rubin, Robert Herman, François Schweizer, and W. Kent Ford, Jr., for wise comments on the manuscript.

Dark Halos Around
Spiral Galaxies

1984

When I place the slit of a spectrograph along the image of the major axis of a spiral galaxy viewed at reasonably high inclination to the line-of-sight, and expose for several hours, the spectrogram that results resembles that shown in Fig. 1. The strongest emission line is from Hα. From its measured displacement, I map the velocity at successive distances from the nucleus. Even to the eye, it is apparent that velocities are not falling at large nuclear distances, as is expected for Keplerian orbits about a massive distant nucleus. Orbital velocities which are constant with increasing distance imply that the interior mass is increasing linearly with r, and that the local mass density is falling with r as $1/r^2$. Surface brightness in the disk is observed to fall more rapidly, i.e., nominally exponentially with r. This observation leads to the inference that the mass per unit luminosity is increasing with radius; the increase in mass is presumed due to the presence of non-luminous matter distributed at large radial distances. Stellar orbital velocities remain high in response to the gravitational attraction of the unseen mass.

I show this fundamental observational material for two reasons, both slightly philosophical. Astronomers and physicists face very different problems in performing their experiments, and I hope that physicists would be interested in seeing the raw astronomical observations. Equally important, I would like to impress you with the ease of this experiment in contrast to the complicated particle experiments that physicists are carrying out. While I can forget to open the dark slide or place the plate in the hypo before the developer, as long as I make no mistakes, valuable results are obtained.

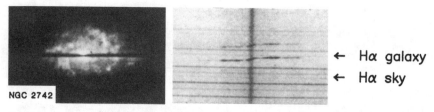

← Hα galaxy
← Hα sky

Figure 1. *The spiral galaxy NGC 2742 as seen reflected on the slit jaws and viewed on the TV monitor at the 4-meter telescope of the Kitt Peak National Observatory, Arizona, USA. The dark line crossing the galaxy is the slit of the spectrograph. Light falling through the slit is dispersed into the spectrum shown at right, obtained with the aid of an image tube. The red spectral region is to the top of the spectrum, the blue is to the bottom. The horizontal lines crossing the spectrum arise from OH in the earth's atmosphere, except for the marked atmospheric Hα. The separation of this line from the Hα line in the galaxy measures the redshift of the galaxy. Due to the rotation of the galaxy, the left side is approaching the observer, and the right side is receding, with respect to the nuclear velocity. These latter emission lines are shifted toward the red.*

These observations reveal the presence of significant mass out to the edge of the optical image, but they do not define the distribution of this mass, nor the extent of the mass beyond the optical disk. I show you now one example where nature offers us the opportunity to probe a galaxy potential well beyond the optical disk. A0136-0801 (Fig. 2) is one member of a class which now contains about a dozen galaxies. It consists of a flattened rotating disk, with a thin ring running almost over its pole. Its disk nature is revealed by its surface brightness profile; its rotation is indicated by a strong stellar velocity gradient along the principle plane, and lack of velocity gradient along the minor axis. The ring contains normal galactic matter; gas, stars, and dust, as revealed by its knotty appearance, emission line spectrum, and deep absorption where it passes in front of the disk. Matter in the ring is rotating, but its axis of rotation lies almost in the plane of the disk.

François Schweizer, Brad Whitmore and I [1] have used the velocities in the disk and in the ring to probe the potential of the system. The radial extent of the ring is about three times that of the disk. In the ring, orbital velocities remain high, equal to velocities at the limits of the disk. Hence, we infer that the mass continues to rise linearly with radius to a distance three times the radius of the disk. Moreover, the match of the velocities in the vertical and horizontal planes tells us that the shape of the potential is more nearly spherical than flattened.

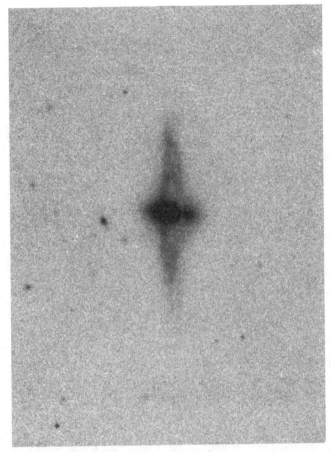

Figure 2. *A0136-0801, from a Cerro Tololo 4-meter plate by F. Schweizer.*

Discussions of precession and stability of rings about galaxies [2] like A0136-0801 convince us that the formation of this galaxy was a two-step process, most likely involving the capture of matter by the existing disk galaxy. But regardless of the past history, this curious object permits us to detect directly for a single galaxy what had only previously been inferred: a spheroidal halo of dark matter is situated beyond the optical disk and contributes a significant—probably the major—amount of the galaxy mass.

References

1. F. Schweizer, B. C. Whitmore, and V. C. Rubin, Astron. J. **88**, 909 (1983).
2. J. E. Tohline, F. F. Simonson, and N. Caldwell, Astrophys. J. **252**, 92 (1982).

How Much Dark Matter Is There?

1989

H
ow much of the matter in the universe is dark? Let's start by discussing how much matter we can see. When you attempt to build a galaxy out of stars and gas and dust, as Beatrice Tinsley and Richard Larson and others have done, you find that the mean of the ratio of mass-to-luminosity, *M/L*, is of the order of 1 or 2. In a typical galaxy, for each unit of mass there is, on the average, one or two units of luminosity measured in units of solar mass and solar luminosity. Thus, from their visible spectral properties, most spirals appear to have an average stellar population not too unlike that of the Sun. But the galaxy dynamics tell a different story. Across the optical image of a galaxy, the luminous disk is matched by an equal nonluminous halo mass. Moreover, beyond the optical galaxy out to the largest radii to which rotation velocities have been measured, the dark matter amounts to five or ten times the luminous matter.

To pursue the answer for a larger region of the universe, we must enter the realm of cosmology. In cosmology, observational facts are hard to come by. However, most astronomers do agree on a few that are essential to any theory describing the history and evolution of galaxies. These are:

1. The universe originated in a Big Bang, whose laws of physics—at least after the first few minutes—are the same laws of physics we know today.
2. The 3 degrees Kelvin black body radiation is the remnant of the primordial fireball: it has been expanding and cooling ever since. Due to the expansion of the universe, the distance between isolated galaxies is presently increasing. Only within the dense clusters do the mutual gravitational attractions of the galaxies keep these regions from participating in the universal expansion.
3. Our galaxy (and hence the observations which we make from the

Earth) is not at rest with respect to this remnant radiation. But once this motion is accounted for, the background radiation is isotropic on small angular scales to an impressively high degree, a few parts in 100 000. This is generally interpreted to mean that the early universe was smooth and isotropic.

4. The region over which we observe luminous galaxies is clumpy on enormous scales. Regions of long strings of superclusters separate large regions void of any luminous galaxies. Observers have not yet examined any large volume of space in which the distribution of galaxies is smooth.

In addition, we must define a few terms in order to evaluate the mean density of luminous and dark matter. We accept the following consistent set of values where H_0, the value for the expansion rate of the universe, is uncertain within a factor of 2, but most of the other listed quantities will scale for other values of H_0; G is the gravitational constant; $M_\odot$ is the solar mass; $L_\odot$ is the solar luminosity; and ρ_0 is the mean mass density.

$H_0 = 50$ kilometers per second per megaparsec

$\rho_0 = 3\Omega\, H_0^2/(8\,\pi G)$

$M = \rho_0 \sim \Omega\, 10^{11} M_\odot$ per cubic megaparsec

$L \sim < 10^8\, L_\odot$ per cubic megaparsec

$\Omega = \rho_0/\rho_c$

$M/L \simeq \Omega\, 1000\, M_\odot/L_\odot$

If there is sufficient mass in the universe, the mutual gravitational attraction of all the mass will ultimately halt the expansion; it could even cause the universe to start contracting. The mean mass density which will *just* halt the expansion, i.e., close the universe, is designated ρ_c (where the c stands for closure or critical). Ω is the ratio of the observed density ρ_0 to the closure density ρ_c. The mean mass density of the universe, calculated from gravitational theory, is of order $\Omega\, 10^{11}\, M_\odot$ Mpc^{-3}, or $1000\, \Omega$ times larger than the mean luminosity density, $\langle L \rangle$. The mean luminosity density is a very uncertain number, for it comes from counting up the luminosity from all the faint galaxies, an uncertain guess at best, and it also depends upon the dimming effects of the gas and dust within our galaxy. But as these adopted values show, it will take a value of $M/L \simeq 1000$ to close the universe. Thus, on average, there must be 1000 solar units of mass for each solar unit of luminosity. This density corresponds to placing a galaxy 1/10 the mass of the Milky Way in each cubic megaparsec of the universe. But such high densities are not seen optically—nor even from the dynamical studies described above. The dynamical results imply values of M/L of tens and hundreds, but not as high as 1000. Order-of-magnitude results for various dynamical systems are shown in Table 1. While exact numbers are controversial, I think this set of values is a fair assessment of the current status of our knowledge.

As is apparent from the upper part of the table, various dynamical analyses give values of M/L less than about 200, corresponding to a value of $\Omega < 0.2$. Thus, all of the dynamical studies would be satisfied with a universe whose mass is only about 1/5 the critical mass, but one in which the quantity of dark matter exceeds by factors of 10 the luminous matter.

A measure of the density of the universe comes also from the theory of nucleogenesis in the conventional Big Bang cosmology. Because deuterium, produced in the initial Big Bang, can only be destroyed by subsequent evolution, its current abundance is a measure of the density of the universe. Early studies by Tinsley, David Schramm, and coworkers, and recent detailed analysis by Jean Audouze and his colleagues, are all consistent in deducing a low universe density; i.e., $\Omega < 0.2$. Thus, our universe can be one in which the dark matter is 'normal'; that is, the type that makes up galaxies, stars, and atoms. This matter, termed baryonic, has been processed in stars and has evolved along with the universe. Such a

TABLE 1. *Values of M/L and Ω for various dynamical systems.*

System	Mass	Scale (kiloparsecs)	M/L	Ω
Visible				
Galaxy	10^9–10^{11}	25	2	0.002
Dynamical				
Galaxy	10^{10}–10^{12}	25	10	0.01–0.02
Binary	10^{10}–10^{13}	50	50	0.05
Group	10^{13}	150	150	0.15
Cluster	10^{14}	2000	250	0.25
Local Supercluster	10^{15}	20 000	300	0.15–0.3
Deuterium abundance				<0.2
Inflation				1
Closed				>1

universe is both consistent with all of the dynamical observations and will expand forever.

However, such an 'open' universe poses questions which theorists find difficult to answer. How did galaxies initially form in such a universe? Why is the microwave background radiation so smooth, given the lumpy distribution of matter in a universe so large that regions of it could never have communicated with each other? And, why is $\Omega = 0.2$ so tantalizingly close to $\Omega = 1$? Why, for example, is Ω observed not either orders of magnitude smaller or larger than unity?

Given these valid questions, theorists have postulated a universe in which $\Omega = 1$. In such a universe, the amount of dark matter is five times larger than that required by the dynamical arguments, and by the limits imposed by nucleogenesis, this dark matter cannot be baryonic. As Alan Guth explains, these $\Omega = 1$ models modify conventional Big Bang cosmology to incorporate a time of enormously rapid inflation during the initial universe, thus solving the smoothness problem. Such a universe will not expand forever, but will slowly coast to a halt. Only if $\Omega > 1$ will the universe ultimately contract.

We may simplify the current ideas concerning models of the universe into these two extremes: one with $\Omega = 0.2$, which is derived from dynamical arguments, and one with $\Omega = 1$, which is derived from theoretical arguments. There is, however, still a third model, which is presently favored by only a very few. In this model, 'what you see is what you get.' That is, the distribution of mass in a galaxy is described by the distribution of light, but Newtonian potential theory is assumed to be not valid for the very low values of acceleration encountered. In this model, Newton's simple law cannot be the correct form for analyzing the distribution of mass in a galaxy; rather, the observed distribution of light gives rise to the observed velocities. Consequently, there is no invisible matter. The mean density of the universe is just that which is visible, and is therefore very small, $\Omega = 0.002$ or so. Most astronomers prefer to accept a universe filled with dark matter rather than to alter Newtonian gravitational theory. Yet Mordehai Milgrom and Jacob Bekenstein, two proponents of a modified theory, point out that laws of physics have been altered in the past on the basis of evidence weaker than the evidence that has led astronomers and physicists to postulate the existence of invisible matter.

We conclude with an uncertainty: there is significantly more dark matter than luminous matter in the universe. But whether that dark matter comprises 90% or 99% of mass is unknown; whether the universe will continue to expand forever is unknown. Even Dennis the Menace (Fig. 1) recognized the problems in determining how much dark matter there is.

DENNIS the MENACE

"LOTS OF THINGS ARE INVISIBLE, BUT WE DON'T
KNOW HOW MANY BECAUSE WE CAN'T **SEE** THEM."

*Figure 1. DENNIS the MENACE® used by permission of Hank Ketcham and ©
by North America Syndicate.*

What Is the Dark Matter?

What is the dark matter made of? Answers to this question contain
many 'ifs'. If we live in a low-density, $\Omega \sim 0.2$ universe, it could all be
baryonic; that is, the normal stuff that makes stars and galaxies and us.
And, if it is baryonic, it may be faint brown dwarfs, small stars which
never became hot enough to start conventional nuclear processing. Or a
large population of undetected white dwarfs. Or enormous numbers of
cold planet-like objects. Or mini-black holes. Or even maxi-black holes—
remnants of the early universe.

If the dark matter is not baryonic—not formed from neutrons and protons—then the large-scale distribution of galaxies in the universe is a clue
as to its nature. If galaxies formed in a universe of exotic particles, these
particles determined the dynamical evolution of the early universe, and by
extension, the nature of the universe we inhabit today. A signature of this
early universe must still be present in the large-scale distribution of galaxies which we currently observe.

The plot of the million brightest galaxies (Fig. 1, p. 8), illustrates a complex distribution of chains, strings, filaments, and voids, all connected in a lace-like pattern. This plot is visual evidence of the nonrandom distribution of bright galaxies. But not only is the two-dimensional distribution of galaxies clumped; velocities are also clumped. Radial velocity measures of galaxies observed beyond the constellation of Bootes by Kirshner, Oemler, Schechter, and Shectman offered impressive evidence that there is a region void of galaxies, which extends over 160 million light years. These observations, coupled with the identification of other large regions void of galaxies, have produced an enormous industry in N-body calculations, as astronomers attempt to identify the physical and dynamical conditions which would give rise to such clumpy distributions. The deepest survey to date of galaxy distribution is that of Huchra, Geller, and de Lapparent. They obtained velocities for all galaxies brighter than magnitude 15.5 in a strip of sky 6° by 120° going through the Coma Cluster. Assuming that the Hubble flow is smooth, an assumption which has yet to be seriously tested by independent distances to galaxies at large distances, these results show the galaxies arranged around the peripheries of giant voids, forming structures as large as the scale of the observations. These enormous features may be coming close to violating the smoothness observed in the microwave background. They certainly tax standard gravitational models. They may instead imply explosive hydrodynamical formation, as Ostriker and Cowie suggested some years ago. They also allow the possibility that structure size may continue to grow with sample size; we may not yet have observed a large enough fraction of the universe to comprehend its structure.

Currently, we do not even know if anything *is* present in the voids. Are voids empty of all matter? Or only void of baryonic matter? Are low-luminosity galaxies present there? Is there a critical density, below which galaxies will not form? Does the formation of some galaxies inhibit the formation of others? These are some of the questions which follow from the apparently clumpy distribution of the most luminous galaxies.

Nonbaryonic forms of matter that have been suggested include neutrinos, gravitinos, photinos, sneutrinos, axions, magnetic monopoles, and dozens more. The cosmology predicted by each of these particles is complex, and none of them has properties which predict all of the observations. Neutrinos were one of the first appealing candidates for the dark matter. Neutrinos are known to exist; they have generally been assumed to have zero mass. Ramanath Cowsik and John McClelland pointed out in 1973 that if the neutrino had a mass in the range of tens of electron volts, this would have interesting cosmological implications. In the years since then, experimental evidence for the existence of a neutrino mass has been questioned. Most recently, the detection on Earth of neutrino events which arose from the supernova in the Large Magellanic Cloud has reopened the discussion of the neutrino's mass.

But neutrinos are hot, relativistically moving particles, and a universe dominated by neutrinos would form enormous structures early in its history. Such a universe would produce a 'top-down' cosmology, in which the largest structures form first, substructures form clusters next, and galaxies form last. Such a cosmology has interesting large-scale properties, which match well the clusters and superclusters we see as long strings in the plots of galaxy distribution. Despite this promise, the initial enthusiasm for a neutrino-dominated universe has faded. First, it is still not established that the neutrino has mass. Second, fragmentation of the largest structures down to the sizes of galaxies would take an appreciable fraction of the age of the universe and galaxies would have only recently formed. This timescale seems incompatible with our present ideas of galaxy evolution.

Hence, many cosmologists now favor an alternative model, one in which the particles dominating the universe are cold rather than hot. Such particles, of which the axion is an oft-chosen one, have never been detected, but they are allowed by the physics of the theories from which they emerge. Early in such a universe, the cold axions form clouds which withstand the expansion and clump on all sizes, from clusters of stars to clusters of galaxies. Because fluctuations on all scales condense at the same time, many of the curious interactions between galaxies and their environment can be understood. The overwhelming drawback of such models is that they are based on particles whose existence is currently only postulated. Laboratory experiments presently underway are attempting to detect the axion. Particle physics and astronomy will grow even closer with a detection of such exotic particles.

Those of us curious about matter in the universe hope for better answers to the questions: What is it? And how much is there? Whatever it is, it must be dark, it must clump about galaxies, it must be less concentrated to the centers of galaxies than is the light, and it must not appreciably obscure the background galaxies. Astronomers are fond of saying that the dark matter could be cold planets, dead stars, bricks, or baseball bats. Physicists are fond of saying that it could be billions of mini-black holes, or somewhat fewer maxi-black holes, or, indeed, any one of a number of exotic particles from the zoo of objects permitted by physical theories but never yet observed. Whatever it is—and it could be of more than one type—it must be the major constituent of our universe.

In a very real sense, astronomy begins anew. The joy and fun of understanding the universe we bequeath to our grandchildren—and to their grandchildren. With over 90% of the matter in the universe still to play with, even the sky will not be the limit.

A Century of Galaxy Spectroscopy[a]

1995

Introduction

V irtually all of our knowledge concerning galaxies has come during the twentieth century. From the work of Russell, Payne-Gaposchkin, Oort, Morgan, Hubble, Shapley, and others, we know that we live in a galaxy, that the solar system orbits about the distant Galactic center, that galaxies are expanding one from the other. But Hubble's concept of Island Universes evolving in splendid isolation has given way to a dynamic universe. Astronomers now recognize that a galaxy is a continuously evolving structure that will acquire stars or lose stars through gravitational interactions; it will acquire gas or lose gas through infall or galactic winds; it will be actively forming stars or quiescent depending upon its most recent history, and it will look like an elliptical or a spiral depending upon the eyes of the beholder, the limiting magnitude of the telescope exposure, and the spectral band. Just as continents are moving and evolving beneath our feet, so too galaxies are assembling and transforming over our heads.

My family is fond of a Peanuts cartoon in which Lucy is asking, "On the good Ship of Life, Charlie Brown, which way are you going to place your deck chair? To see where you have been or to see where you are going?" And Charlie Brown replies, "I can't seem to get my chair unfolded." Well, my chair is unfolded, and I am going to discuss our increasing understanding of galaxy kinematics over the last century. Certainly the last century is not more important nor more interesting than the next. In

[a] Russell Prize Lecture, delivered January 9, 1995 at the 185th meeting of the AAS in Tucson, Ariz.

fact, we are all aware that we are living at the start of a remarkable era during which our knowledge of the universe will escalate, due to new telescopes, new detectors, new techniques, new computers, and especially new ideas. However, it is clear that every major problem of extragalactic astronomy that we have attacked during the twentieth century is still unsolved; unknown are the amount and distribution of dark matter, the rate of expansion of the universe, the magnitude and significance of large scale motions, the detailed motions of gas and stars within galaxies.

The Early Years

One-hundred years ago, no one knew what a galaxy was. But 96 years ago, Scheiner [1] reported the first successful spectrum of a galaxy, in a two-page paper in Volume IX of the *Astrophysical Journal*, "On the Spectrum of the Great Nebula in Andromeda":

> The continuous spectrum can be clearly recognized on it from F to H, and faint traces extend far into the ultraviolet. A comparison of this spectrum with a solar spectrum taken with the same apparatus disclosed a surprising agreement of the two, even in respect to the relative intensities of the separate spectra regions ... No traces of bright nebular lines are present, so that the interstellar space in the Andromeda nebula, just as in our stellar system, is not appreciably occupied by gaseous matter.

Scheiner had discovered that M31 is a stellar assemblage. Scheiner did not publish the spectrum, and over the years I repeatedly asked astronomers from Potsdam Observatory to attempt to locate it. In 1987 Hans Oleak wrote me that he had found the plate (incredible, he said, after two major wars and numerous upheavals), traced it, and was to publish it [2]. A handwritten note on the plate envelope said the plate had been chemically enhanced to increase its intensity, but the enhancement had disappeared in 1906.

By 1912, Slipher at the Lowell Observatory was obtaining spectra of galaxies—objects he thought to be planetary systems in formation. By 1914 he had confirmed that M31 and NGC 4594, the Sombrero, exhibited inclined lines [3,4]. Inclined spectral lines were familiar to astronomers because of their knowledge of planetary rotation. A spectrum obtained along the equator of Jupiter shows each line to be inclined. Light from one limb is blue shifted and from the other limb is redshifted with respect to the center, as the planet's rotation carries one limb toward and one away from the observer. Hence, inclined lines in a galaxy spectrum were evidence that galaxies are rotating.

Within a few years, Pease confirmed that M31 was rotating [5]. Using the Mt. Wilson 60-inch, he exposed a spectrum along the minor axis dur ing 84 hours in August, September, and October 1916 and along the major axis for a total of 79 hours in August, September, and October 1917. The lack of significant velocity gradient on the minor axis, and the steep velocity gradient over the inner 2.5 arcminutes along the major axis were evidence of rotation. It is curious, and interesting, that Pease took the minor axis spectrum before the major axis one. Pease apparently chose to make the first exposure along the minor axis, knowing that inclined lines were observed by Slipher along the major axis. A misprint in Pease's paper lists the wavelength interval as λ4930 (rather than the correct 3930) to λ4950. Dividing by the 5.3 mm extent, Rubin and D'Odorico noted an erroneous dispersion [6]. N.U. Mayall and R. Minkowski both wrote to me to point out the error, with Minkowski stating, "It would have been a miracle if Pease had been able to observe M31 with 3.8 Å/mm."

The twenties was a decade of rapidly increasing understanding of galaxies. Using the velocity curve of M31 from Pease, Oepik published "An Estimate of the Distance of the Andromeda Nebulae," which I consider to be one of the most original papers of this century [7]. By now, the *Ap.J.* had abstracts, and his begins:

Andromeda Nebula.—Assuming the centripetal acceleration at a distance r from the center is equal to the gravitational acceleration due to the mass inside the sphere of radius r, an expression is derived for the *absolute distance* in terms of the linear speed v_o at an angular distance ρ from the center, the apparent luminosity i, and E, the energy radiated per unit mass. From observations, v_o comes out 15 km/sec. for $\rho = 150$ arcseconds; and giving i a value corresponding to 6.1, and assuming E the same as for our galaxy, the distance is computed to be 450,000 parsecs. This result is in agreement with that obtained by several independent methods. If it is correct, the *mass* within 150 arcseconds is 4.5×10^9 times the sun's mass, and the nebulae is a stellar universe comparable with our galaxy.

Oepik determined a distance for M31 such that when angular distances within M31 were transformed to linear distances, the mass-to-luminosity ratio (in today's notation) was equal to that of the solar neighborhood, a quantity Oepik had also to derive. His analysis indicated that M31 is a massive distant galaxy. But Oepik was too modest. The derived 450 kpc distance is closer to the 700 kpc used today than the 210 to 250 kpc distances that Hubble and other astronomers were using until the 1950s.

In 1927, Oort explained observed stellar motions within our galaxy as the result of a differentially rotating galaxy, with angular velocity decreasing with nuclear distance [8]. By 1929, velocities for 46 galaxies were available [9], all but a few of them due to Slipher. Slipher deserves more credit than he generally gets in histories of astronomy. With velocities available for 46 galaxies but distances for only 18, Hubble [9] determined the motion of the galaxy and the expansion rate of the universe, over 500 km sec^{-1} Mpc^{-1}. This result, more credible than earlier solutions (e.g. Ref. 10), was enshrined in the literature by Hubble's enormously influential book, "The Realm of the Nebulae," from which my generation of students learned about galaxies [11]. Rarely mentioned is Oort's analysis of the same data, but with two original differences [12]. Distances were measured in units of A, the (then uncertain) distance to M31. The expansion velocity he found is L = 140 km/sec.A; with the present M31 distance this corresponds to ~200 km sec^{-1} Mpc^{-1}. But forward thinking, Oort concludes his paper:

> In order to see whether the factor with which the recession increases with distance is the same in different parts of the sky the nebulae may suitably be divided into two groups: those south of the galactic plane and those north of it. Assuming the solar velocity as found in solution (A) I find from the 13 objects south of the galactic plane L = 142 km/sec.A ± 21 (m.e.) The whole material gave +140 km/sec.A; L is thus found to be practically the same for northern and southern nebulae, that is for two regions of the sky which are separated by about 120°.

Possibly without intending to, Oort had founded the study of large-scale motions in the universe. The next discussion of galaxy velocities as a function of position on the sky would apparently wait 20 years [13].

Spectrographs were slow, exposures were long, and velocities were uncertain due to instabilities in the instruments, low dispersions, and low signal-to-noise ratios. For his thesis, Babcock determined velocities of rotation within M31 using the Lick Observatory Crossley telescope, observing diffuse nebulosities near the nucleus, and 4 HII regions located 30 arcminutes to 100 arcminutes distant [14]. Exposure times were as great as 20 hours. Velocities, which continued to rise with distance rather than to fall as expected from a Keplerian inverse square law, were imprecise but still revealed a bit of the truth. The apparent solid body rotation pattern beyond the nuclear region puzzled theorists, but was neatly explained in terms of Newtonian gravitational theory by Holmberg [15].

Mayall continued the observation of rotation velocities within M31, from 31 HII regions (Fig. 1) [16]. Exposure times with the Crossley had decreased to about 10 hours; a distance for M31 of 230 kpc was assumed. A comparison of this rotation curve with that known for our Galaxy, convinced Mayall that something was wrong. He ended:

> Conclusion: It thus appears that either (1) the main bodies of M31 and the Galaxy are appreciably dissimilar in size, or (2) the accepted distance of M31 is too small, or (3) the present simplified analysis of Cepheid radial velocities is misleading. Whether any of these three alternatives, separately or in combination, is correct probably cannot be decided until more precise data are available for distances of Cepheids in the Galaxy and in M31.

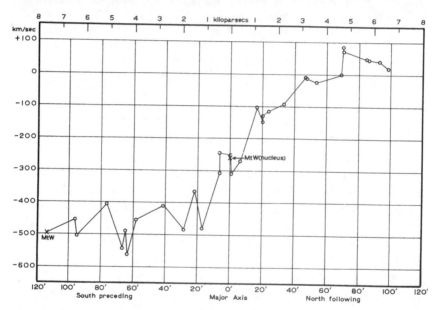

Figure 1. *Observed velocities in M31, from Mayall [16]. The HII regions with measured velocities are indicated on the image. Note that some indication of the complex nuclear velocities are indicated in these early observations.*

By mid-century, radial velocities for about 800 galaxies were available from Humason, Mayall, and Sandage [17]. A value of H = 180 km sec^{-1} Mpc^{-1} was derived. Other questions, such as evidence of large scale structure and large scale motions, were still to be asked. But even for those questions which had been asked, answers were not imminent. The distance scale was still in question. The forms of rotation curves, dozens compiled by Margaret and Geoffrey Burbidge and generally covering only the inner parts of galaxies, were assumed to be falling and free of complication [18]. Very few voices were raised in query. One of these few was Oort's; his presentation at the dedication of the McDonald observatory discussed the distribution of light and the distribution of matter in NGC 3115 [19]. He wrote in the abstract:

> It is found that beyond about 10 arcseconds from the center in the equatorial plane the mass density must be about constant. The actual density depends on the unknown flattening of the attracting mass, but it should be at least of the order of 140 solar masses per cubic parsec (10^{-20}gm/cm^3), or 2000 times that near the sun. In this region of constant mass density the light-density diminishes with a factor of at least 10: the distribution of mass in the system appears to bear almost no relation to that of the light.

This statement is the earliest statement I know which identifies the necessity for dark matter in individual galaxies. Thirty years later Freeman raised a similar caution for NGC 300, now with better velocity information available [20]:

> The HI rotation curve has V_{max} at R ~15 arcminutes, which also happens to be the photometric outer edge of the system. If the HI curve is correct, then there must be undetected matter beyond the optical extent of NGC 300; its mass must be at least of the same order of magnitude as the mass of the detected galaxy.

Yet it is important to remember that these few sentences were lost in the mass of publications which "detected" Keplerian falling velocities. The unexpected relation between luminosity and velocity was noted by very few, even though Ostriker and Peebles had suggested a halo as a means of stabilizing a galaxy disk [21].

During the 1970s accurate rotation velocities in galaxies accrued rapidly, due principally to more sophisticated observing instrumentation and techniques, both in the optical and in the radio spectral regions. By 1975 there was an optical rotation curve for M31 (Fig. 2) that was flat beyond the nucleus and believable, with accurate optical velocities [22] which extend to 120 arcminutes, matching precise radio velocities which extend to 170 arcminutes [23]. High resolution optical spectra across the nucleus still took 6 hours to obtain, even using a spectrograph which incorporated an image tube. The telescope operators at the 84-inch KPNO telescope would cringe when they saw us coming, for they knew it meant a long night of only 2 exposures, with the telescope operator sitting on the observing platform as required, with no job to perform after once setting the telescope at the start of the night. Telescope operators even covered the faintly luminous clock face, for we feared red light leaks in the spectrograph during the long exposures.

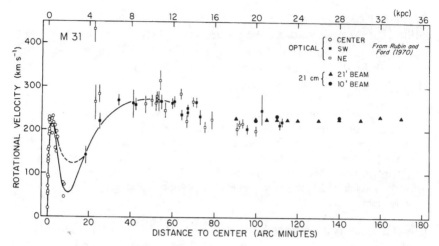

Figure 2. *Rotation velocities in M31 as a function of distance from the nucleus. The optical velocities come from Rubin and Ford [22]; the 21 cm velocities from Roberts and Whitehurst [23].*

But there was still more to learn. Brinks' thesis, a high resolution study of HI velocities in the inner 60 arcminutes of M31, showed additional complexity; along many lines of sight a low velocity component coexists with the higher rotational velocity [24]. While generally interpreted as outer disk gas warped back into the line-of-sight, and hence with only a small radial component of velocity, we should be aware that bars or other triaxial motions near nuclei may be responsible (e.g., Ref. 25). As in M31, ionized gas with low velocities coexists with the higher rotation velocities in many spirals I observe.

But lest you think that M31 was the first galaxy to define a flat rotation curve, let me refer you to the 1959 study by Louise Volders, from HI observations made at Dwingeloo [26]. For M33, HI velocities, as if drawn with a ruler, slice through a scatter of optical points. The lack of impact on the astronomical community is curious; perhaps the instrumental capability was doubted. Surely a falling rotation curve was expected, based on the rotation curve for our galaxy predicted by Oort's constants A and B (and Oort was Volder's professor). By 1972 Rogstad and Shostak [27] could assemble 5 flat HI velocity curves for Scd galaxies, obtained with the Owens Valley two-element interferometer. They conclude their abstract:

Because of the very flat rotation curves observed for these Scd galaxies, total masses extrapolated to infinite radius are not known. Surface mass-luminosity ratios must reach values of ≈20 at the Holmberg radius.

Before that decade was over, we all knew that rotation curves for spiral galaxies are flat (e.g., Ref. 28), and the efforts to determine the distributions and amounts of the mass components are still underway (e.g., Ref. 29). For every galaxy, we can determine only the lower limit to its mass, for a flat rotation curve implies that the mass increases linearly with radius and is not converging to a limiting mass, a distinctly untasteful circumstance in physics. Consequently, M/L increases with radius, as mass M continues to increase where optical luminosity L has already converged to its limiting value. Hence 21-cm studies of spirals with large HI disks are especially valuable for inferring the distributions and amounts of dark matter at large radial distances.

Our Galaxy is no exception. The mass interior to the sun's distance is $10^{11} M_{\odot}$. Mass continues to rise almost linearly with increasing nuclear distance, even beyond the distance of the LMC. This result comes from a study of the space motion of the LMC [30]:

> ... the Galactic halo has a mass ~$5.5 \pm 1 \times 10^{11}$ M$_{\odot}$ within 100 kpc and a substantial fraction (1/2) of this mass is distributed beyond the present Galactic distance of the Magellanic Clouds (≥ 50 kpc). This mass is nearly half that assumed in the previous models, but it is consistent with some recent estimates for the galactic halo. Beyond 100 kpc this mass may continue to increase to ~10^{12} M$_{\odot}$ within its tidal radius (~300 kpc).

Other techniques for deriving mass also indicate that the galaxy mass continues to grow to over $10^{12} M_{\odot}$ at 200 or 300 kpc [31–33]. This distance is getting suspiciously close to the halfway point between our galaxy and M31. Occasionally I am glad that the matter is dark, or we might find it difficult to be optical astronomers in the brighter universe.

What is the dark matter, especially now that J. Bahcall and colleagues [34] have not detected red dwarfs? Big Bang nucleosynthesis [35] tells us that there must be dark baryons, even for a low density universe; I'm optimistic that we will find them. Their detection will occupy astronomers of the next century. But my record for predictions is poor—in 1980 I said we would know what the dark matter is by 1990, from work by particle physicists. Perhaps what I call the Lalande principle will come into play, and knowledge will come from an unexpected direction. In 1779, Lalande said we would not know the solar motion for several hundred years, until the sun had moved into a new region of stars. Four years later, Herschel determined the solar motion from the proper motions of 12 stars.

Finally, note that the alternative to dark matter, a modification of Newtonian gravitational theory (e.g., Refs. 36 and 37) now seems less likely. X-ray observations [38,39] reveal that the geometrical distribution of luminous matter differs from that of dark matter in clusters and in some galaxies. In flattened clusters of galaxies, the dark matter distribution is rounder than the luminous matter distribution. Hence the luminous galaxy distribution traces neither the gravitational potential nor the gravitating matter. Unless dark matter in clusters is distinctly different from dark matter in galaxies, a modified form of Newton's laws which accounts for galaxy rotation curves will not explain the distribution of matter in clusters.

Partly Peculiar Galaxies

According to Struve, if you pick five stars at random, one of them will be peculiar [40]. Well, if you pick and study one galaxy at random, it will be peculiar. For my random galaxy, I pick M31. An Hα image (Fig. 3) of the central region of M31 [41] constructed by Ciardullo from almost 400 individual Hα frames, exhibits an aspect ratio close to face-on, unlike that of the outer M31 disk which is more nearly edge-on. In addition, there is a marked asymmetry between the circularity of the S and N arms. As Fig. 3 shows, this morphology is a good match to nuclear gas velocities measured 25 years ago [42] which reveal both a nearly circular morphology, and a N/S difference. Recently, Stark and Binney [43] have reproduced the major features, using a triaxial model:

> ... the ionized gas distribution (and to a lesser extent the dust distribution) in the inner few arcminutes of M31 forms a spiral pattern with nearly circular symmetry on the plane of the sky ... this pattern occurs in the transition region where the gas is shocked as it moves from the inner x_1 orbits to the outer x_2 orbits. The spiral looks round because the x_1 orbits are elongated along the line of sight ...

Thus, the nucleus of M31 offers complexities we only now start to interpret, complexities not only in the gas. Close to the nucleus, Kormendy [44] observes stars, with a steep gradient of velocity with radius, and with a large velocity dispersion. This combination is diagnostic of the presence of a supermassive nuclear black hole of mass 10^7 or $10^8 M_\odot$. And within 0.5 arcseconds the HST detects not one but two nuclei [45,46]; one bright, one fainter. Lauer *et al.* write:

> V- and I-band *HST Planetary Camera* images of the great spiral galaxy in Andromeda, M31, show that its inner nucleus consists of two components separated by 0.49 arcseconds. The outer isophotes of the nucleus at 1.4 arcseconds $\leq r \leq 3.0$ arcseconds are elongated, but are concentric with the M31 bulge. The nuclear component with the lower surface brightness, P2, is also coincident with the bulge photocenter ... The brighter nuclear component, P1, is well resolved and corresponds to the nuclear core imaged by *Stratoscope II*.

They suggest that the brighter nuclear region may contain its own black hole, whose presence keeps the knot from being torn apart. The *Stratoscope*

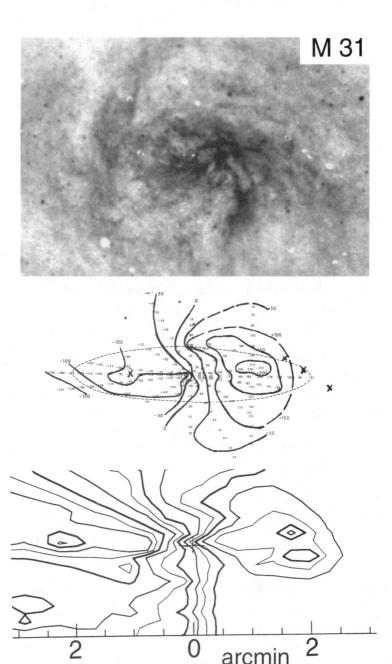

Figure 3. (Top) Hα plus [NII] negative image of inner region of M31 [41]. Note the fine threads of emission curving outward, the more face-on appearance than the optical disk, and the asymmetry. (Middle) Line-of-sight velocities from ionized gas in M31 [42]. Ellipse represents a thin disk 400 parsecs (2 arcminutes) viewed at an angle of 77°. Contour intervals 50 km/sec. (Lower) Isovelocity contours from model [43], intervals 30 km/sec. Note the general agreement between the observations and the model. The lower scale applies to all images. Two minutes of arc is only 2% of the M31 radius.

II reference is to the early [47] balloon flight which imaged the core of M31. Understanding the kinematics and the past and future evolution of the nucleus of M31 will offer us much to ponder.

Multi-Spin Galaxies

Only during the past thirty years have astronomers been able to determine the kinematics of more complex galaxies, the pathological ones. Many of us who had earlier studied spiral galaxies finally had the observational instrumentation, the reduction facilities, and the theoretical framework with which to attack these observationally difficult objects. Just as in medicine, where the study of rare pathological cases often leads to an improved understanding of more normal specimens, the observation of extragalactic systems with tidal distortions, polar rings, and kinematically discrete cores has emphasized the important role that gravitational interactions play in the evolution of all galaxies.

Georgia O'Keefe wrote "It takes time to see, just as it takes time to make a friend." Arp [48] was one who took the time to see, and in his beautiful *Atlas of Irregular Galaxies*, he arranged a wide variety of pathological types into sequences by morphological appearance. The volume is a monument to the importance of just looking, even before an understanding of process is in place. Only now do we begin to understand the geometry and the kinematics of the interactions which produced the sequences he noted.

NGC 5128 (Cen A), an apparent elliptical with prominent distorted dust lanes across its minor axis, had long been a puzzle to astronomers. In his volume, *Galaxies*, Shapley [49] wrote:

> NGC 5128 is a 'pathologic specimen'—one of the external galaxies with a peculiar spectrum. Any fully successful theory of structure must take into account such abnormal forms.

and questioned if NGC 5128 was the result of a merger. Burbidge and Burbidge [50] heroically observed it (declination = −46°) from McDonald (latitude = +30°), and detected a rotation in the disk component. But it was Ulrich's [51] discovery that the filaments wrapping the minor axis of the Helix galaxy, NGC 2685 (see also Ref. 52), were rotating in a plane parallel to the galaxy minor axis that alerted astronomers and initiated a new direction in galaxy kinematics. Shortly thereafter, Graham [53] showed that the filaments in NGC 5128 comprise a disk in rotation:

> The present observations reinforce the view that NGC 5128 is a giant elliptical galaxy in which is embedded an inclined and rotating disk composed partly of gas. The disk also contains stars and could be similar to a small galaxy ... The structural

peculiarities ... suggest some sort of massive addition of gaseous material to a basi-
cally normal elliptical galaxy in the not too distant past.

Within a few years, Schweizer, Rubin, and Whitmore [54] measured for
a polar ring galaxy, A0136-0801, the rotation of both the lenticular galaxy
and the ring encircling its poles. These curious objects are important because (1) they make it possible to infer the three-dimensional potential of
the galaxy and, hence, to learn about the distribution of dark matter out of
the principal plane; (2) they emphasize the importance of a "second event"
in the lives of these galaxies. In each case, the second axis of rotation must
have been established by gas collected after the initial rotation axis had
been established.

Polar ring galaxies are now understood as galaxies which acquired cold,
infalling gas (e.g., Ref. 55) which smears out over several orbital periods
into a ring. A ring orthogonal to the principal galaxy is stable over many
Hubble times; a ring intermediate between pole and disk precesses to the
disk [56,57] and ultimately settles into a stable orbit where it may start
forming stars. Both observations and computer simulations reveal that the
orbits of this gas may be prograde or retrograde, or even skew with respect
to the orbits of the preexisting stars.

The knowledge of polar ring galaxies and their derivatives lessened somewhat the surprise of finding significant numbers of elliptical and S0 galaxies with small disks of cold gas. These disks, with higher angular momentum than could arise from gas shed by evolving stars, are presumably of
external origin, perhaps even infallen tidal debris. Their kinematics, generally decoupled from the larger galaxy, are complex and sometimes
triaxial; their location is generally in the core but sometimes more extended. A compendium of E and S0 galaxies with nuclear gas disks [58]
counted 9 with counterrotating gas, 3 with skew rotation, 13 with corotating
gas and stars.

In some cases, the nuclear gas disks have aged and formed stars. Nuclear
stellar disks too may orbit prograde, retrograde, or skew to an existing
stellar disk [59,60]. And in yet other cases, the cold gas is external to the E
galaxy, as in IC 2006. Here, an HI ring rotates counter to the rotation of the
elliptical [61] offering the opportunity to determine the shape of the potential in the plane (circular: [62]) and the mass beyond the optical galaxy
(continuing to rise, with mass-to-luminosity ratio increasing by a factor of
about 4).

And in an even more novel development, radial velocities of planetary
nebulae, at large radial distances, are now used to study the kinematics of
complex objects. For NGC 5128, velocities of over 400 planetaries show
rotation in two coordinates, along the dust lane, and perpendicular to it,

i.e., along the major axis of the elliptical [63,64]. The mass-to-light ratio increases with nuclear distance, from 3 at origin to 10 at 20 kpc, offering evidence for dark matter. HI too has now been detected [65] in the outer region, some perhaps related to the blue knots detected optically [66].

So a surprise of the 1980s was that we could learn about motions within ellipticals, once thought to be gas free, by study of their gas components, and even by their planetaries. Thirty years earlier, Osterbrock [67] had studied ionized gas in a few ellipticals, but the field could not flourish until sophisticated observing techniques were widely available. For spheroidal galaxies, a rotating disk in equilibrium with the overall gravitational potential presently offers the best opportunity for determining the mass distribution.

Some observers of nuclear disks sit not at a telescope but at a computer. Fifteen years of computer simulations of two merging disks of "stars" developed ultimately to simulations which included also "gas" particles [68,69] and demonstrate the important phenomena which occur when even a small amount of gas is included. Calculations show that when two disks merge, tidal torques cause the interacting gas to lose angular momentum, to be funneled efficiently to the center where the resulting angular momentum of the gas disk is unrelated to the initial angular momentum. Statistics of merging disks containing gas [70] show that the inclusion of as little as 1.5% gas makes the remnant rounder, less triaxial, and hence more in keeping with the observations than with theoretical predictions of extreme triaxiality.

Thus, computer models successfully approximate some of the properties of spheroidals with gas disks. Observers too have now initiated attempts to date ancient mergers. Schweizer and Seitzer [71] have shown that the residuals in the broad-band color-magnitude relation correlate with morphological evidence of the merger: ripples, tidal tails, and isophotal peculiarities. Both morphology and spectroscopy reflect the age since the merger. A more comprehensive summary of these multi-spin galaxies is given elsewhere [72]. A more artistic summary is given in the 1963 painting of Remedios Varo, "Phenomenon of Weightlessness" (see Plate 1); [73]. Here, the professor stands amazed, each foot in a different sense of gravity, as a dark mass outside the window pulls an astronomical model into the second frame. This painting offers a moving, artistic view of galaxies with two senses of rotation.

So the end of the eighties left us confident: theory, observations, and computer simulations could explain much of the curious kinematics that our improved instrumentation and techniques could uncover. We did not predict the larger surprises that were shortly to come.

The Decade of the Nineties: Seeing the Future

With the decade of the nineties only half gone, we already glimpse some future directions of galaxy spectroscopy, and I offer four examples. Other features, not yet imaginable, must await the future.

NGC 4826

While multiple axes of rotation could be understood for hot spheroidals, it was not expected that colder spirals would show equivalent complexity. But early in the decade, Braun, Walterbos and Kennicutt [74,75] uncovered a surprise. Neutral hydrogen in the outer disk of the Sab galaxy NGC 4826 (Fig. 4; the Sleeping Beauty Galaxy, nee the Black Eye) was counterrotating. From earlier optical spectroscopy [76], the sense of rotation of the inner excited gas was known: from early radio observations, it had gone unnoticed that the outer HI was counterrotating. Recent optical observations [77,78] detail motions in the inner region, including the prominent dusty lane, where the stars and gas orbit in concert and in the transition region where the gas undergoes an orderly, rapid fall from 180 km/sec prograde to 200 km/sec retrograde, along with an infall of 100 km/sec. However, the stars continue their prograde rotation.

NGC 4826 can be understood as the equatorial counterpart of a polar ring galaxy, in which gas, captured in an oblique angle and smeared into a ring, precesses to the plane where its sense of rotation is counter to that of stars and gas in the preexisting disk. However, because the acquisition of only a few percent of the galaxy mass is expected to heat up and destroy the disk [79] it is still unclear how nature manages this delicate feat of maintaining the disk.

NGC 4550

NGC 4550 (E7/S0) remains the only galaxy known with two coincident, coplanar stellar disks, one disk rotating prograde, one retrograde. A gas disk is coincident with one of the stellar disks. Details of its discovery and its kinematics are published elsewhere [80–82]. Although such a configuration has a distinguished mathematical history (e.g., Refs. 83 and 84), it had been assumed to be unrealistic, for no mechanism for its formation could be imagined. Toomre's [84] first sentence,

> The galaxy models about to be displayed here belong perhaps only in the category of elegant curiosities.

is no longer true. Like NGC 4826 and other multi-spin galaxies, these forms spring from the merger of a stellar disk and a counterrotating gas

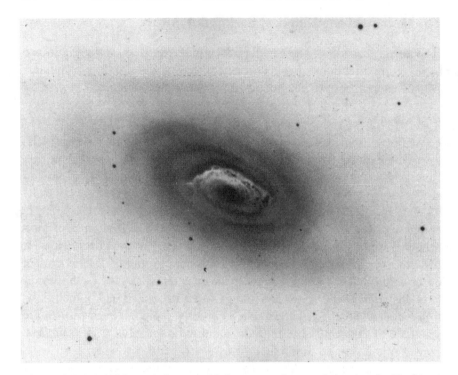

Figure 4. *NGC 4826, copied at very high contrast from a plate taken by Ritchie at the Mt. Wilson 60-inch telescope over the nights of May 5,6,7,8, 1910, and kindly made available by A. Sandage and J. Bedke. From the center outward through the dusty region (here shown light), the stars and gas rotate in the same sense. Beyond the dust lane, the gas reverses direction, and counterrotates with respect to the stars.*

mass which ultimately settles to the disk and forms stars. We may some-day know how galaxy evolution with retrograde capture differs from galaxy evolution which involves prograde capture.

Located near the center of the Virgo cluster, NGC 4550 is undistinguished by its morphology. Recent HST nuclear images show arcs of absorbing material extending into the center, in one location obscuring a bit of the small (~1 arcsecond) central disk, prominent in the V–I image. Bryan Miller, Brad Whitmore and I are now starting to examine these high spatial resolution images.

NGC 4550 and NGC 4826 have taught us that not only elliptical galaxies, but spirals too encompass a wider range of kinematical complexity than had been previously imagined or believed realistic. Such galaxies cause us to enlarge the domain of "what is possible." They also force us to

Plate 1. *Phenomenon of Weightlessness, by Remedios Varo, 1963. An astronomy professor stands with feet placed in two different senses of gravity, while an earth/moon model leaves the horizontal plane and aligns with the inclined sense. This artist's image is a fine illustration for multi-spin galaxies, galaxies with multiple orbital planes, or counter-rotating components. (Reproduced with the permission of Walter Gruen.)*

enlarge the reduction procedures for measuring gas and stellar velocities. The usual computer reduction programs which pick the peak of a Gaussian line profile are not capable of revealing that spectral lines in NGC 4550 consist of two separate components. New reduction techniques [85–88] determine, for example, the detailed velocity pattern of the stars from the integrated line of sight velocity distribution at each radial distance. Using such a procedure, Merrifield and Kuijken [87] have established that 30% of the stars in NGC 7217 form a distinct component which orbits retrograde.

M87

We had long anticipated faint object spectroscopy (FOS) from the repaired Hubble Space Telescope, and our expectations were rewarded with a view of the kinematics near the nucleus of M87, the peculiar elliptical near the center of the Virgo cluster. From the abstract of "*HST* FOS Spectroscopy of M87: Evidence for a Disk of Ionized Gas Around a Massive Black Hole," Harms *et al.* [89], write:

> Radial velocities of the ionized gas in the two positions 0.25 arcseconds on either side of the nucleus are measured to be ~ ± 500 km s^{-1} relative to the M87 systemic velocity. These velocities plus emission-line spectra obtained at two additional locations near the nucleus show the ionized gas to be in Keplerian rotation about a mass M = $(2.4 \pm 0.7) \times 10^9$ M$_\odot$ within the inner 0.25 arcseconds of M87.

The broad emission wings on the nuclear spectrum suggest even more rapid rotation within 0.25 arcseconds (18 pc); by the time this paper is published, spectra obtained with the 0.09 arcseconds aperture may have yielded even tighter limits on the volume encompassing the central black hole.

NGC 4258

As anticipated, HST has obtained spectra of galaxy nuclei at significantly higher resolution than possible from the ground. But suddenly, on a very special object, ground bound astronomers have observed the nucleus of a galaxy at a resolution several orders of magnitude higher than that of HST. The object is NGC 4258, the technique is VLBA, and the sources are water vapor masers [90,91]. These authors observe numerous water vapor masers situated in a ring, orbiting the nucleus. Inner and outer dimensions are 4 and 8 milli-arcseconds (mas; 4 mas = 0.13 pc). The velocities imply a steep rise to 1000 km/sec, then fall in an exact Keplerian pattern (Fig. 5). Note that velocities near 500 km/sec come from masers distributed along the inner edge of the ring, all approximately equidistant from the center. The observed gradient arises from the varying projection along the line-of-sight. A mass of 3.6×10^7 M$_\odot$ in a region less than 0.13 pc (27,000 AU) is implied. The mass density, $\geq 4 \times 10^9$ M$_\odot$ pc^{-3}, exceeds by a factor of 40 that for any previous black-hole candidate, and is thus compelling evidence for a massive central black hole.

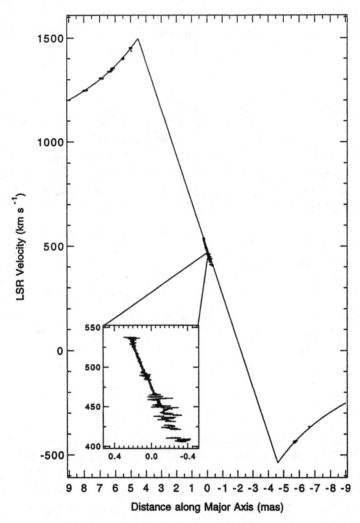

Figure 5. *Velocities of water vapor masers in the nuclear ring of NGC 4258 from Very Long Baseline Array (VLBA) observations [90]. At the distance of NGC 4258, 4 mas (milli-arcseconds) = 0.13 parsecs = 27000 AU. One AU is the distance of the sun from the earth; we are seeing details on sizes approaching solar system dimensions in a distant galaxy. The solid line shows the line-of-sight ring velocities, followed by a Keplerian ($r^{-1/2}$) fall. A mass of 3.6×10^7M$_\odot$ is interior to 0.13 parsecs.*

Concluding Random Thoughts

Virtually all of the research programs attacked by spectroscopic observations of galaxies during the past century are in need of additional study. These include: the amount, distribution, and composition of dark matter; the rate of the expansion of the universe; the character of the distribution of galaxies; the amount, cause, and significance of large scale motions; the ages and evolution of galaxies, the structure and motions at very small nuclear distances; the prevalence of supermassive nuclear black holes.

This impressive list offers much for observers to pursue. Yet surely the role of observers is to confound the theorists. Thus when the *New York Times* [92] headline read "New Surveys Of the Universe Confound Theorists" to describe the wedge velocity plot [93] of thousands of galaxies, I felt pretty smug. As an observer, I was not confounded, and I was pleased that the theorists were. However, when I displayed the printed page at Durham, England, Carlos Frenk pointed out the headline of an adjacent, unrelated article, "The Brain May See What Eyes Cannot." Touché, Carlos. The combination of brain and eye will resolve these outstanding questions.

As an extragalactic observer, my livelihood has been based upon the existence of the Doppler shift. I hope other extragalactic observers too recognize the Doppler shift as a gift of nature, especially because its magnitude does not diminish with increasing distance. But it has occasionally concerned me that we live in a radially expanding universe, when radial motions are the only component that we can measure. So I wonder if there will be surprises in the transverse components, once we can measure sufficient transverse velocities at large distances.

Finally, I stress once again my belief that many deep mysteries of the universe are yet to be discovered. In a spiral galaxy, the ratio of dark-to-light matter is about a factor of 10. That's probably a good number for the ratio of our ignorance-to-knowledge. We're out of kindergarten, but only in about the third grade. A good third grade summary might be the Monty Python song, "The Galaxy Song." We have learned a lot. We will have much joy in discovering some of the rest.

Acknowledgments

I thank the many colleagues and post doctoral fellows who have contributed so actively to my understanding of the history and practice of galaxy kinematics. They have been a continued source of information, ideas, and encouragement, especially Drs. R. J. Rubin, W. K. Ford, Jr., E. M. Burbidge, G. R. Burbidge, D. Burstein, R. Ciardullo, S. D'Odorico, J. A. Eder, J. A. Graham, D. A. Hunter, J. P. D. Kenney, B. Miller, C. J.

Peterson, M. S. Roberts, F. Schweizer, N. Thonnard, and B. C. Whitmore. I thank D. E. Osterbrock for his careful and cheerful comments on the manuscript. I am grateful to many observatories for making telescope time available: Cerro Tololo, Kitt Peak, Las Campanas, Lowell, and Palomar Observatories. I also happily acknowledge the active support from the Carnegie Institution of Washington.

References

1. J. Scheiner, Ap.J. **IX**, 149 (1899); also A.N. **148**, 325 (1899).
2. W. Bronkalla and H. Oleak, in Proceedings of the 10th European Regional Astronomy Meeting of the IAU (Prague, 1987).
3. V. M. Slipher, Lowell Obs. Bull. No. 62, (1914).
4. M. Wolf, Vierteljahresschr. Astron. Ges. **49**, 162 (1914).
5. F. G. Pease, PNAS **4**, 21 (1918).
6. V. C. Rubin and S. D'Odorico, A&A **2**, 484 (1969).
7. E. Oepik, Ap. J. **LV**, 406 (1922).
8. J. H. Oort, BAN **4**, 275 (1927).
9. E. Hubble, PNAS **15**, 169 (1929).
10. C. Wirtz, Scientia **38**, 303 (1925).
11. E. Hubble, *The Realm of the Nebulae* (Yale University Press, New Haven, 1936).
12. J. H. Oort, BAN **5**, 235 (1930).
13. V. C. Rubin, AJ **56**, 47 (1951).
14. H. W. Babcock, Lick Obs. Bull **19**, 41 (1939).
15. E. Holmberg, MNRAS **99**, 650 (1939).
16. N. U. Mayall, in *The Structure of the Galaxy* (University of Michigan Press, Ann Arbor, 1951), p. 19.
17. M. L. Humason, N. U. Mayall, and A. R. Sandage, AJ **61**, 97 (1956).
18. E. M. Burbidge and G. R. Burbidge, in *Galaxies and the Universe*, edited by A. Sandage, M. Sandage, and J. Kristian, *Stars and Stellar Systems IV* (Univ. Chicago Press, Chicago, 1975), p. 81.
19. J. H. Oort, Ap. J. **91**, 273 (1940).
20. K. C. Freeman, Ap.J. **160**, 811 (1970).
21. J. P. Ostriker and P. J. E. Peebles, Ap. J. **186**, 467 (1973).
22. V. C. Rubin and W. K. Ford, Jr., Ap. J. **159**, 379 (1970).
23. M. S. Roberts and R. N. Whitehurst, Ap. J. **201**, 327 (1975).
24. E. Brinks, Ph.D. thesis, University of Leiden (1984).
25. M. R. Merrifield and K. Kuijken, BAAS **26**, 1493 (1995).
26. L. Volders, BAN **14**, 323 (1959).
27. D. H. Rogstad and G. S. Shostak, Ap. J. **176**, 315 (1992).
28. V. C. Rubin, D. Burstein, W. K. Ford, Jr., and N. Thonnard, Ap. J. **289**, 81 (1985).
29. S. M. Kent, AJ **96**, 514 (1988).
30. D. N. C. Lin, B. F. Jones, and A. R. Klemola, Ap. J. **439**, 652 (1995).
31. P. J. E. Peebles, Ap. J. **362**, 1 (1990).
32. A. S. Kulessa and D. Lynden-Bell, MNRAS **255**, 105 (1992).
33. D. Zaritski, R. Smith, C. Frenk, and S. D. M. White, Ap. J. **405**, 464 (1993).
34. J. N. Bahcall, C. Flynn, A. Gould, and S. Kirhakos, Ap. J. **435**, L51 (1994).

35. C. J. Copi, D. N. Schramm, and M. S. Turner, Science **267**, 192 (1995).
36. M. Milgrom, Ap. J. **333**, 689 (1989).
37. J.D. Bekenstein and M. Milgrom, Ap. J. **286**, 7 (1984).
38. D.A. Buote and C. R. Canizares, Ap. J. **400**, 385 (1992).
39. D. A. Buote and C. R. Canizares, Ap. J. **427**, 86 (1994).
40. C. H. Payne-Gaposchkin, AJ **82**, 665 (1977).
41. R. Ciardullo, V. C. Rubin, G. H. Jacoby, H. C. Ford, and W. K. Ford, Jr., AJ **95**, 438 (1988).
42. V. C. Rubin and W. K. Ford, Jr., Ap. J. **170**, 25 (1971).
43. A. A. Stark and J. Binney, Ap. J. **426**, L31 (1944).
44. J. Kormendy, Ap. J. **325**, 128 (1988).
45. T. R. Lauer *et al.*, AJ **106**, 1436 (1993).
46. R. M. Rich, K. J. Mighell, and W. L. Freedman, BAAS **26**, 1434 (1995).
47. E. S. Light, R. E. Danielson, and M. Schwarzschild, Ap. J. **194**, 257 (1974).
48. H. Arp, *Atlas of Peculiar Galaxies* (Calif. Inst. of Technology, Pasadena, 1966).
49. H. Shapley, *Galaxies* (The Blakiston Company, Philadelphia, 1943).
50. E. M. Burbidge and G. R. Burbidge, Ap. J. **129**, 271 (1959).
51. M.-H. Ulrich, PASP **87**, 965 (1975).
52. P. L. Schechter and J. E. Gunn, AJ **83**, 1360 (1978).
53. J. A. Graham, Ap. J. **232**, 60 (1979).
54. F. Schweizer, B. C. Whitmore, and V. C. Rubin, AJ **88**, 909 (1983).
55. E. O. Athanassoula and A. Bosma, ARA&A **23**, 147 (1985).
56. J. E. Tohline, G. F. Simonson, and N. Caldwell, Ap. J. **252**, 92 (1982).
57. T. Y. Steiman-Cameron and R. H. Durisen, Ap. J. **263**, L63 (1982).
58. F. Bertola, L. M. Buson, and W. W. Zeilinger, Ap. J. **401**, L79 (1992).
59. R. Jedrzejewski and P. Schechter, AJ **98**, 147 (1989).
60. M. Franx, G. Illingworth, and T. Heckman, Ap. J. **344**, 613 (1990).
61. F. Schweizer, G. H. van Gorkom, and P. Seitzer, Ap. J. **338**, 770 (1989).
62. M. Franx, J. H. van Gorkom, and T. de Zeeuw, Ap. J. **436**, 642 (1994).
63. X. Hui, PASP **105**, 1011 (1993).
64. X. Hui, H. C. Ford, R. Ciardullo, and G. H. Jacoby, ApJS, **88**, 423 (1993).
65. D. Schiminovich, J. H. van Gorkom, J. M. van der Hulst, and S. Kasow, Ap. J. **423**, L101 (1994).
66. J. A. Graham, BAAS **26**,1504 (1994).
67. D. E. Osterbrock, Ap. J. **132**, 325 (1960).
68. L. E. Hernquist and J. E. Barnes, Nature **354**, 210 (1991).
69. J. E. Barnes and L. E. Hernquist, Ap. J. **370**, L65 (1991).
70. J. E. Barnes, in *The Formation of Galaxies*, (to be published) (Canary Island Winter School, 1994).
71. F. Schweizer and P. Seitzer, AJ **104**, 1039 (1992).
72. V. C. Rubin, AJ **108**, 456 (1994).
73. J. A. Kaplan, *Unexpected Journeys: The Art and Life of Remedios Varo.* (Abbeville Press, New York, 1988).
74. R. Braun, R. A. M. Walterbos, and R. C. Kennicutt, Jr., Nature **360**, 442 (1992).
75. R. Braun, R. A. M. Walterbos, R. C. Kennicutt, Jr., and L. J. Tacconi, Ap. J. **420**, 558 (1994).
76. V. C. Rubin, E. M. Burbidge, G. R. Burbidge, and K. H. Prendergast, Ap. J. **141**, 885 (1965).
77. V. C. Rubin, AJ **107**, 173 (1994).

78. R. A. M. Walterbos *et al.*, AJ **107**, 184 (1994).

79. G. Toth and J. P. Ostriker, Ap. J. **389**, 5 (1992).

80. V. C. Rubin, J. A. Graham, and J. P. D. Kenney, Ap. J. **394**, L9 (1992).

81. H.-W. Rix, M. Franx, D. Fisher, and G. Illingworth, Ap. J. **400**, L5 (1992).

82. V. C. Rubin, Mercury **XXII**, 109 (1993).

83. D. Lynden-Bell, MNRAS **120**, 204 (1960).

84. A. Toomre, Ap. J. **259**, 535 (1982).

85. O. E. Gerhard, in *ESO/EIPC Workshop on Structure, Dynamics, and Chemical Evolution of Early-type Galaxies*, edited by J. Danziger, W. W. Zeilinger, and K. Kajar (ESO, Munich, 1993), p. 331.

86. H.-W. Rix and S. D. M. White, MNRAS **254**, 389 (1992).

87. M. R. Merrifield and K. Kuijken, Ap. J. **432**, 575 (1994).

88. R. P. van der Marel and M. Franx, Ap. J. **407**, 525 (1993).

89. R.J. Harms *et al.*, Ap. J. **435**, L31 (1994).

90. H. Miyoshi, J. Moran, J. Herrnstein, L. Greenhill, N. Nakai, P. Diamond, and M. Inoue, Nature **373**, 127 (1995).

91. L. J. Greenhill, J. M. Jiang, J. M. Moran, M. J. Reid, K. Y. Lo, and M. J. Claussen, Ap. J. **440**, 619 (1995).

92. J. N. Wilford, *New York Times* (January 15, 1991), C1.

93. V. de Lapparent, M. G. Geller, and J. Huchra, Ap. J. **302**, L1 (1986).

Part IV
THE ASTRONOMICAL LIFE: WOMEN IN SCIENCE AND OTHER HEROES, COLLEAGUES, AND FRIENDS

S cientists are individuals with their individual styles of doing science. Warm friendships are made and become an important part of a life in science. Like other disciplines, science has its own sociology and an historical record that has discouraged women and minorities from entering the field. I believe that science profits from diversity and that the joy of doing science should be open to all. Thus, I have continually encouraged young women to study science in spite of the discouragements they may receive along the way. Progress in enlarging the fraction of women and minorities is slow, perhaps glacial, but the gradient is now positive.

"An Unconventional Career" was an interview with Sally Stevens, an editor of *Mercury*, 1992.

"Women's Work" was originally a lunch talk given at the headquarters of the American Association for the Advancement of Science (AAAS) in recognition of Women's History Month, at the invitation of Shirley Malcom. I was not planning to publish it, but Shirley was persuasive. It appeared in *Science* 86. Publications by Margaret Rossiter and Deborah Warner have taught me about the history of women in science.

"Opening the Doors" was one of two after-dinner talks given at the conference on "Recruitment and Retention of Women in Physics," Chevy Chase, Maryland, 1990; sponsored by the American Institute of Physics.

The other talk was given by our daughter, astronomer Dr. Judith Young. It never was published, although parts were excerpted in *Physics Today*, 1992.

"Sofia Kovalevskaia: Scientist, Writer, Revolutionary" was a review published in 1985 in *The Mathematical Intelligencer* (Springer–Verlag: NY).

"George Gamow" is the introduction to a lecture I delivered at a summer school in Varenna, Italy in 1982. It was published in *Gamow Cosmology*, LXXXVI Corso, Soc. Italiana di Fisica, Bologna, 1986.

"E. Margaret Burbidge" was published in *Science*, 1981, on the occasion of her election to the presidency of the AAAS. Margaret is an exceptional astronomer who has served as role model for many women astronomers.

"The Leviathan of Lord Rosse" is the introduction to the Grubb–Parsons Lecture which I gave at the University of Durham, England in 1993. It is published here for the first time. The iron works of William Parson, the third Earl of Rosse, ultimately became the Grubb–Parsons Company, builders of telescopes.

"Gérard de Vaucouleurs and the Local Supercluster" was published in 1989 in *Gérard and Antoinette de Vaucouleurs: A Life for Astronomy* (edited by M. Capaccioli and M. G. Corwin, Jr., World Scientific, Singapore), a volume in honor of the 70th birthday of Gérard. This essay introduced Gérard's reprinted 1957 paper containing early ideas of the Local Supercluster.

"Ralph A. Alpher and Robert Herman" have been family friends ever since 1951 when Bob Rubin shared an office with Ralph at the Applied Physics Laboratory of the Johns Hopkins University, and Bob Herman's office was down the hall. It was their collaborations with George Gamow which led me to write my Ph.D. thesis under Gamow's direction. I was privileged to give this brief statement when Alpher and Herman were awarded the Henry Draper Medal of the National Academy of Sciences in 1993.

"Desert Island Disks" is a Public Radio production of WETA, Washington D.C., which airs weekly. It is a descendent of the original British program of the same name, in which the participant is both interviewed and chooses five musical selections (which are played on the program) to accompany her on a desert island. The program aired in 1992.

"Difficult Questions" is an answer to one of the very many letters I get from school children; this 1992 query was one of the best.

"Where in the World is Berkeley, California?" is a commencement talk given to the Astronomy, Physics, and Physical Science graduates at the University of California, Berkeley, in May 1996.

An Unconventional Career

1992

E arly in December 1991, Vera Rubin spoke about her life and work with Mercury Assistant Editor Sally Stephens. Their conversation ranged from Rubin's youth, through the experience of being the first woman to (legally) use the telescopes on Mount Palomar, to her pioneering work establishing the presence of vast quantities of "dark matter" in the universe.

Sally Stephens: You knew from the time you were quite young that you wanted to do astronomy.

Vera Rubin: Yes, from about the age of 10, I knew I wanted to be an astronomer. I didn't know a single astronomer, male or female. Being a woman didn't bother me at all. I didn't think that all astronomers were male, because I didn't know. I didn't even know how you became an astronomer.

Q: You went to Vassar College.

A: I needed a scholarship and they gave me one. And I knew Vassar taught astronomy. I knew about the work of Maria Mitchell. So I went to Vassar and had a very, very nice classical education in astronomy and really learned a lot. It's not the way astronomy is taught nowadays. But I really got a lot out of it.

Q: In what way was it different?

A: Well, I did a lot of classical dynamics. I worked through Smart's books on galactic dynamics and stellar astronomy, calculating orbits on a desk calculator. I learned a lot of what we now would probably call practical astronomy, coordinate systems, and mathematical descriptions, more than astrophysics.

Q: After graduating from Vassar in 1948, you started graduate school at Cornell.

A: I went to Cornell because I had gotten married and my husband was doing his Ph.D. there. At the time, Martha Stahr Carpenter was one of two people in the Cornell Astronomy Department; the other one was a navy navigator. She had just recently gotten a degree at Berkeley and she was very interested in galaxy dynamics. So I continued doing some galaxy dynamics with her. There were then radial velocities known for about 108 or 109 galaxies. For my master's thesis under her I asked the question—if you removed the Hubble expansion of the galaxies, did they have any other systematic motion?

I remember she wrote to Milton Humason on the West Coast—this was 1949 or 1950. It was known among astronomers that Humason and his co-workers had many more redshifts which had not been published. He responded by saying they would be public pretty soon [in fact, it was six years before the redshifts were published]. So I was sort of told by observers that I should wait until there were more redshifts. And it was known that Princeton's Kurt Gödel was working on models of rotating universes and such, and so I was sort of told by theorists that I should wait. I guess I was impatient and so I went ahead with the paper on the galaxies.

Q: You did find some systematic motions independent of the Hubble expansion.

A: I put every galaxy at the distance its apparent magnitude would put it. I took out the motion the expansion of the universe would contribute at that distance. I then took what was left—what we call the residual motion—and just plotted them on a globe. I presume none of this work would hold up today. I think the magnitudes of the galaxies were not good enough, and the velocities were probably not good enough.

I found that many of these galaxies defined a great circle on the sky, or roughly a circle, and that there were large regions of positive and negative values of residual velocity. What in fact I really found was the supergalactic plane, although I entitled the paper "Rotation of the Universe."

Q: But that was not well received?

A: The paper was rejected by both the *Astrophysical Journal* and the *Astronomical Journal*, but I did present it at an AAS meeting in Haverford in December 1950.

Q: But you were vindicated later, weren't you? You found the same lopsided expansion in the 1970s when you were working with Kent Ford.

A: Well in principle, yes. The concept was a meaningful scientific question, as recent work by others on large-scale motions has indicated. [The observation that many local galaxies are being pulled in the direction of Pegasus is now called the *Rubin–Ford effect*.]

Q: Then you went to Georgetown University.

A: We moved to Washington because my husband took a job there, and so I finished my graduate work at Georgetown.

Q: You were going to graduate school, which is hard enough, and you were also married and starting a family. How did you do it?

A: It was the hardest thing in my life. I was at Cornell two or three years. I knew I wasn't going to be there long enough to get a Ph.D., so I didn't enter a Ph.D. program, but I did a lot of physics. I studied physics under Philip Morrison, Richard Feynman and Hans Bethe. I did a lot of coursework. I had our first child just as I was finishing that. I actually took my master's orals just a couple of weeks before I had the baby.

So I had a masters and I knew galactic dynamics and I knew some physics. And then I entered Georgetown to do a [Ph.D.] thesis and I got connected with George Gamow, who was a professional colleague of my husband's. Ralph Alpher and Bob Herman, who had been working with Gamow on Big Bang cosmology, were also colleagues of my husband's. Gamow asked me a very interesting question and I decided to try to answer it as a Ph.D. thesis. The question was whether there was a scale length in the distribution of galaxies. And I got hold of Harvard galaxy counts on the sky and I applied a two-point correlation analysis to it. So my entire Ph.D. work really consisted of a single correlation analysis, something you would now do on a computer in minutes. Although in those days it was a messy job getting the data together. In the late 1960s, the Japanese applied correlation techniques to the distribution of galaxies, and then Jim Peebles started in the 1970s.

Q: There are a lot a women astronomers who are trying to juggle family and career. You've been successful both in your career and with your family. Do you have any advice for them?

A: Well, I guess the only advice I would have is just sort of muddle through. It gets easier. The most important thing of course, and I was truly blessed, is to marry the right man. And I don't say that lightly because it's very easy for a marriage to suffer and even not to survive, if one is really, *really* dedicated to anything, including astronomy. I had the advantage that my husband was as dedicated to his physics as I was to my astronomy, but that we also both were enormously interested in having a family and raising the children. For many, many years, our lives consisted of nothing but our research and our family. The children really benefited because they had a very active father who helped raise them. And it's not all disadvantages.

If you are very, very busy, you don't have to pretend you aren't. I always needed help, and so running the house was always everybody's job. I was very outspoken when I was raising the children, informing them as often as necessary that I was not the maid. And we also had the great fortune of living in a very large old house in Washington so everybody had lots of space; that really helped. The kids had their rooms and I never cared what they looked like. We have lots of large tables. We have a very large table in the dining room, and another in the breakfast room. My husband used to always humorously say that there would always be a space on one of the tables where we could eat. Things were always spread out on those tables. I did a lot of my work at home. So having a family and career was very hard, but it's do-able.

Q: You were the first woman to be permitted to observe at Palomar?

A: I was the first woman to be *legally* permitted to observe at Palomar. Margaret Burbidge had observed there. Geoff Burbidge, Margaret's husband, a theorist, not an observer, had a Carnegie post-doc. He was thus eligible for telescope time and she certainly went up and used the telescope, but she was not permitted to stay there. How often they let her, if it was frequent, I don't know but I suspect it was only a few times. But she had gotten up there, under his name. I was the first to observe there in my own right.

Q: One of the arguments that was made against having woman observers was that there weren't appropriate bathroom facilities.

A: That is correct.

Q: When you observed, was that a problem?

A: No, but let me tell you the full story. I now work for the Carnegie Institution, and I have been reading its yearbooks, starting in 1902 when Andrew Carnegie formed the Carnegie Institution and work on Mt. Wilson Observatory began. He wanted to build a facility, and he says so, where the men would not be bothered by their families. So he called the living quarters on Mt. Wilson the "Monastery," built small rooms, and did not permit families. I certainly am not excusing anybody, but this problem wasn't something that originated in 1948 when Palomar's 200-inch building was built. Even today, there is only one toilet on the ground floor at Palomar and it says "Men."

I was asked by Alan Sandage in 1963–64 if I would like to apply for telescope time at Mt. Wilson or Palomar. I was sent a proposal form which said, printed, "Due to limited facilities, it is not possible to accept applications from women." And someone had penciled in "usually."

My first night on the mountain, it snowed. I had asked for 48-inch Schmidt time and I was given it. Olin Eggen was on the 200-inch. So he took me around the 200-inch, and with great majesty he opened the door and said, "This is the famous toilet." These days they've built a console room off the 200-inch where we work and there's a toilet there and it doesn't say anything on the door. About two years ago, I sort of got annoyed at this "Men" sign and I cut out a little figure with a skirt and pasted it up. It stayed for the four days I was there, but it wasn't there the next year.

Q: Why were you the first? Was it your research that was found to be particularly deserving, did they like you, or was it just that the times had changed?

A: Let me *hope* that the work I was doing was interesting enough. I really had wanted to get back to a program in galaxy dynamics. It was what I knew and it was what I was interested in. And there was some very interesting work going on by Guido Münch who was observing stars in our own galaxy toward the center, trying for the first time to deduce a rotation curve of our own galaxy, star by star. He picked stars and he tried from their spectroscopy to get their distance, and then assumed circular orbits. He had attempted to get rotation curves for the inner part of the galaxy.

My usual interest was sort of orthogonal, or 180°, from others'. I was interested in looking at the outside of the galaxy and trying to deduce the rotation curve for the outer part of the galaxy. I had done some of this with students at Georgetown, just using bright O and B stars from the literature. And ultimately I got to the point where I needed to make my own observations. I just couldn't find distant enough stars. And so in 1963, when Kitt Peak opened, I started using the 36-inch telescope to get orbital velocities of such stars near the anticenter of the galaxy [the direction in the sky directly opposite the Galactic center] to extend the rotation curve of our own galaxy. I had been attempting to get radial velocities of stars as far to the outside of the disc of our galaxy as I could identify stars. That work was very well received; I think people found it interesting. Such a probe of the outside regions of the Milky Way had never really been done before in that fashion and I'd like to think that it was because of the research that I was invited to apply for observing time at Palomar. I clearly had demonstrated that I knew how to use a telescope.

Q: More recently, people have given you a lot of credit for changing the way astronomers look at the universe. They say that 300 years ago astronomers thought that the universe consisted of what they saw. And you came along and all of a sudden the universe really consisted mostly of what you *can't* see. How did all of that come about?

A: This is a very, very long story. It's also really integrated with everything else I'd ever done. It started because I didn't like working on problems that many other people were working on and where I was constantly being besieged with questions about the work. I wanted a problem that I could sit and do at my own pace, where I wouldn't be bothered.

It followed some very early work I had done on QSOs [quasars] with the image-tube spectrograph when I first came to Carnegie. [This was a state-of-the-art instrument (in the 1960s), built by Kent Ford which, like a prism, spread the light from the stars out into their component colors which could then be analyzed. Because it used an electronic enhancer called an image-tube, it was much more sensitive than any earlier spectrographs and could be used to analyze the light from much fainter objects than had previously been possible.] We had an instrument that could look very deep into the universe. We studied quasars for a year or two and I found it personally very distasteful. I just didn't like the pressure of other astronomers calling and asking me if I had observed this and if I knew what the redshift was. I didn't get to a telescope very often and it meant that I either had to give out answers that I was uncertain of, or say I hadn't done it and someone else would then go do it. I just decided that wasn't the way I wanted to do astronomy. I feel that I already have enormous numbers of internal pressures and I don't need external pressures on top of them. I really like to be left alone while I'm working.

So I decided to pick a program that no one would care about while I was doing it. But, at the same time, one that the astronomical community would be very happy with when I was done. So I decided to do a very systematic survey of spiral galaxy rotation curves. [Spiral galaxies are classified according to their appearance. Those with tightly wound spiral arms and a prominent central bulge of stars are called *Sa galaxies*, whereas those with loosely wound spiral arms and a small central bulge are *Sc galaxies*. *Sb galaxies* are intermediate between the two.] My hope was that if I understood the dynamics of spirals, I would begin to understand why spiral galaxies came in different types; that I would learn more about the formation and evolution of spirals by just knowing their dynamics. And so I very carefully set up a fairly long-term program to observe about 60 galaxies, 20 of each kind. I started with the Sc's, of both low- and high-luminosity.

We found something surprising right away, really immediately, probably the first night, with our first few spectra. I did most of the observing with Kent Ford, who had built the spectrograph initially and who is a superb instrumentalist. After our first observing run, we had about a dozen spectra and all the rotation velocities were high, all of them were a surprise.

Q: What is it that you found that was so surprising?

A: Let me tell you what had always been expected. Just following Newton's Law, one expects that as you get farther and farther from a mass concentration, stars' orbital velocities will be slower and slower. This is true in the solar system, where the outer planets orbit much more slowly than do the inner ones.

In the case of galaxies, we were estimating the mass by the luminosity we observed. And because there was so much luminosity in the center, it was assumed that most of the mass was at the center and that the amount of mass would fall off quite rapidly as you went out from the center. So it was expected that ultimately we would get to regions in the galaxy where virtually all of the mass was interior, and the stars' velocities would follow the same pattern as they do in the solar system.

The Burbidges had really done the most beautiful work up to that time on galaxy rotation curves. They were much more limited by their instruments than I was and very seldom did the observations go much beyond the very bright central regions. And you could see the rise and you could see it sort of turn over and flatten out and then everyone just assumed the curve would fall beyond that point. So it was expected that when there was sufficiently sensitive instrumentation to observe far out in the galaxy, you would find that the rotation curve would rise, turn over and fall.

And so I decided to use the gain in sensitivity made possible by the image tube to address this problem. Most people used the image tube to go deeper into the universe—to work on quasars and on the most distant objects. I decided to just go completely across a galaxy and determine the dynamics all the way out. It just was apparent after the first few of these that what we were seeing were almost straight lines on the spectra. You could see it by holding them up.

Q: What did you think when you saw it? This was obviously not what you expected.

A: People keep asking me that. And I have to be really very honest. What I felt initially was just great delight in having gotten to the edge of the galaxy. You first want to see that it's do-able. You devise a program and you try something hard and you look and there's something there. And so what I remember most of all is just this incredible delight at the gorgeous spectra. They really were something! And then I started trying to figure out what was going on and my initial idea was totally wrong. I thought there must be some kind of a feedback mechanism. If a star orbited too rapidly, it was slowed down. And if it orbited too slowly, it was speeded up.

[It turns out that there is no feedback mechanism. The rotation curves did not turn over and fall because there is still a considerable amount of matter in the outer regions of galaxies, regions where few stars are visible. Since we cannot "see" light from this material, it has become known as "dark matter." Astronomers estimate that as much as 90%–95% of all the matter in a galaxy is in the form of this dark matter, and no one is sure exactly what dark matter is. If the estimate is correct, the stars and nebulae we have studied for centuries because we see their light could represent only 5%–10% of what is really there.]

Q: What was the reaction of astronomers to your rotation curves and the existence of dark matter implied by them?

A: I remember that it was not totally accepted by a lot of people. We scientists really change our ideas very slowly. We demand (I think we have to) an enormous amount of evidence before we will change our minds. And I remember some astronomers said, when we published this first dozen very flat rotation curves, "Well, that's because you've observed all the high-luminosity, brighter galaxies. Once you observe the lower luminosities, you'll find falling rotation curves." Actually it turns out that the lower-mass, lower-luminosity galaxies have a larger fraction of dark matter. Their rotation curves never really even turn, they just rise. Even with that evidence, of the first dozen or so rotation curves, it was not overwhelmingly accepted.

Q: What finally convinced people? Was it seeing your spectra with their own eyes?

A: The change came from many directions. I think we still learn a lot from our eyes. And those spectra were so flat, and so really lovely, that you just had to show them to an astronomer and she would understand. Once the concept of a different gravitational potential (one not defined by the light) surfaced, it was kind of shocking. The idea that when we thought we were studying the universe, we were just studying 5% of the universe was amazing. It took a while for everyone to come to terms with the fact that virtually everything we had ever learned in astronomy we had learned through photons. There are a few exceptions—cosmic rays, a few neutrinos, a few things pass by that are not photons. But virtually everything we had ever learned about the universe, we had learned through photons. Suddenly we had to understand that no one ever promised us that all matter would radiate photons—that much of the cosmos could be dark. It does enlarge your vision.

Q: It's interesting because you said that you had wanted to work by yourself on something, and yet your work turned dark matter from speculation into really big news, a hot topic.

A: That's right. Well, you know, we all change and, I have to confess, I have finally gotten into this ... Sometimes I don't pursue directions that might be interesting just because I think they're too controversial. It has been a minor restriction but it really has been my own choice.

Q: Some people have said, when talking about you, that you had a really unconventional career.

A: That's correct. Unconventional certainly in the early part. I did not go to any of the colleges or universities that traditionally turn out astronomers. I didn't study under any astronomers who were research astronomers of note. So I had to learn an enormous amount by myself. And I also was raising a family at the same time. And very happy doing both astronomy and raising a family.

So the kind of life I led was really very different from the kind of life that most graduate students in astronomy were doing. I think it did influence to a very large measure the kind of programs I chose to work on. Had I been at a real establishment institution, I probably would have just absorbed the knowledge that these topics were so speculative, that they were just not worth doing. Or worse than that—the ideas may have just been called wrong and not worth doing. I was not in a position where I was ever told that or exposed to that kind of pressure. So I just went ahead on my own. I think it's also honest to say, that such an approach has a good side and a bad side. Its bad side is that the work was virtually unknown, ignored, or considered wrong. But I think there was at least some intellectual rationale for asking those questions even if they couldn't be answered at that time. It wasn't a bad way to work. I have no regrets.

Q: Is it better for women now than it was when you started out?

A: I think the answer is yes. I'm satisfied that it is improving. But it really is improving very, very slowly. It's partly because university academic departments are filled with people with tenured positions that don't change very often. It will be a long, long, *long* time (unless there is really some very positive action) before astronomy (and physics even more so) is not a man's world. It's going so slowly that at times it's hard to be optimistic.

Q: What do you think it will take to make it better?

A: Well, it will take people wanting it to change. Or it will take a political climate that really believes in affirmative action, that really believes that opportunities ought to be made available equally to everyone. I think at the present time opportunities are not really made equally available. There are always enough men in positions of authority and power who are not enthusiastic about making opportunities available to everyone equally. And a few men in a department who don't want to work with women at all. The opportunities available for women are still fewer than the opportunities available for men.

Q: Why do you think you have been successful as an astronomer—as a woman astronomer—while other women have not and have ended up dropping out.

A: That's not a fair question, because that's the kind of question we only ask someone who has been successful. There is a tragedy in that question and the tragedy is that there may be hundreds and even thousands of women who would have liked to have been astronomers and who would have made great astronomers and really never had the opportunity. They're the ones that ought to be asked what would have been necessary to make it possible for them to do astronomy. Sure, statistically, some people will succeed and maybe it's sheer luck. Maybe I was just at the right place at the right time, and was as good as the males around—as astronomers who were male—so I survived. But I don't know, it's hard to say. I think a good part of it is luck. I think a good part of it is being well enough prepared so you can take advantage of good luck when it comes along. And, for me, I think it's also been having a supportive husband and supportive family.

Q: Your daughter, Dr. Judith Young, is also an astronomer, making you two about the only mother–daughter astronomer pair people can think of. Did she ask for any particular advice or did you give her any particular advice on astronomy as a career?

A: She, all of her high school days, had planned to study biology and chemistry. All of our children went through the Washington, D.C. public high schools. During her senior year in high school, I taught a college-level introductory course in astronomy. I just walked into the high school and volunteered my time. And by the end of that term, she decided she wanted to be an astronomer. I think that all these things she had grown up around just suddenly made sense to her, and must have looked appealing. She has done very well, but still she's had difficulties in her career that would probably not have faced a man at the same level at the same time. She was an undergraduate at Radcliffe, and her very young, very bright advisor suggested she go off and get married the first time she went to him with a problem. She was enrolled as a Ph.D. student when she got married and the faculty decided that she should terminate with a masters because she could not be very serious about being an astronomer if she were getting married. It's hard to say whether—this was in the late 1970s—a young man at the same time would have been hassled because he chose to get married. But she survived even so.

I think most young women astronomers meet an enormous number of hardships during the college days. I think many colleges, universities and graduate schools really still are not totally supportive of young women. Some are, just as some men are. There are some wonderful people in this world who do offer support to all kinds of people. But for those young women students who seem to be having a tough time, my advice would be just not to give up. Not to feel that the problem is with them, but rather to have the confidence to feel that the problem is with the system. And if you can just get through, and get a degree, and get a job, you have a good chance of making it in the astronomical world.

Q: Is astronomy fun?

A: Oh, it's enormous fun. Every day is fun. Observing is spectacularly lovely. Really, really quite inspiring. And I enjoy analyzing the observations, trying to see what you have, trying to understand what you're learning. It's a challenge, but a great deal of fun. There's also this incredible hope that somehow we can learn how the universe works. What keeps me going is this hope and curiosity, this basic curiosity about how the universe works.

Possibly more than many people, I have really been privileged to have a job where I can do very much what I want. And that really means it's such a joy to get up and come to work. Many days I really don't have a very clear idea of what I'm going to be doing. I just sort of continue where the work seems to lead. And that, to me, is just a lovely way to work.

Q: You're lucky to have found something that you enjoy so much.

A: That's correct. I think that probably one of the greatest blessings there is, is to really find every day such a joy.

Women's Work

1986

I was an astronomy student at Vassar College on October 1, 1947, 100 years after the night that Maria Mitchell discovered a comet. Only recently have I realized that no note whatsoever was taken of the centennial of this discovery by the first prominent female astronomer in the United States. Perhaps on that day one of my friends or I irreverently tied a bright scarf around the stern-looking bust of Mitchell that sat in a niche of the observatory building where she taught for many years. But she deserved more.

What I do remember of 1947 is that I wrote a postcard to Princeton University asking for a catalog of the graduate school. Sir Hugh Taylor, the eminent chemist and dean of the graduate school, took the time to answer by writing back that as Princeton did not accept women in the graduate physics and astronomy program, he would not send a catalog. Princeton did not accept women in graduate physics until 1971, in graduate astronomy until 1975, and in graduate math programs until 1976.

For me as a youngster, the account of Mitchell's comet discovery that I found in library books was an exciting part of the lore from the scientific past, along with Benjamin Franklin's kite. Like the kite, it should be a part of every American child's heritage. Yet in 1976, when the Smithsonian Air and Space Museum presented as its first planetarium show a history of 200 years of American astronomy, only male astronomers—all but one of them white—were included. Little boys learned that they could become astronomers. But little girls, who also streamed into the show in enormous numbers, saw that only men were astronomers. After months of effort to have the planetarium show corrected, I received a statement that the talk was recorded and could not be altered.

All of us, men and women alike, need permission to enter and continue in the world of science. In high school and college, students need the permission of parents and teachers. During graduate and postgraduate years, young scientists need the permission of college officials, funding officers, mentors, and colleagues. While such permission has generally been granted to bright men, it has always been less readily granted to young women and continues to be denied to many women even today. In many fields of science, women constitute such a distinct minority—less than five percent of all physicists and seven percent of all astronomers—that they suffer many of the social ills common to minorities.

In Colonial America, public education for women was practically non-existent. But starting about 1820, women's academies came into vogue, and science was a part of the curriculum. By 1871, 18 of these schools had observatories and offered a course in astronomy. Nevertheless, throughout most of the nineteenth century, women in the United States were usually dependent upon a supportive male relative to introduce them to the world of science. For Maria Mitchell it was no different.

The daughter of an intellectual Nantucket family, Maria Mitchell learned from her father how to search the sky with a telescope and how to calculate orbits. Employed during the 1840s and 1850s as the librarian of the Nantucket Athenaeum—the intellectual center of Nantucket and home of literary and philosophical societies, where giants like Thoreau, Agassiz, and Audubon lectured—she studied the advanced astronomical and mathematical texts available to her. Evenings she spent with her father on the roof of their home studying the sky with a telescope. On October 1, 1847, while her parents were downstairs entertaining guests at dinner, the 29-year-old librarian discovered a comet. She promptly announced her discovery to her parents, and Mr. Mitchell immediately posted a note to William Bond, director of the Harvard College Observatory. In 1831, the king of Denmark had offered a gold medal to the next person who discovered a comet with a telescope. (Comets were then generally discovered by eye.) Though the comet was also spotted in Europe, Mitchell's discovery was adjudged to be the first, and the medal was hers.

For the United States, Maria Mitchell became the symbol of women's emergence into the public world of science. In 1848 she became the first woman elected into the American Academy of Arts and Sciences—95 years were to go by until the next woman was admitted. And she was an active member of the American Association for the Advancement of Science. Nevertheless, when Joseph Henry, the first secretary of the Smithsonian Institution, announced in 1848 an "account of a new comet, the discovery of which is one of the finest additions to science ever made in this country," he never identified the "American lady" who made the discovery.

When Vassar Female College opened in 1865, Maria Mitchell was invited to become director of the college observatory and professor of astronomy (Fig. 1). She accepted the positions and remained at Vassar until her retirement in 1888. Like many other women professors then teaching in women's colleges, she had no college education, but she had developed her skills working at the Athenaeum and as a "computer" for the United

Figure 1. *Maria Mitchell (seated), Professor of Astronomy at Vassar College, and her assistant and former student, Mary Whitney, Vassar 1868, in the dome of the Vassar College Observatory. (Photo courtesy of Special Collections, Vassar College Libraries.)*

States Coast Survey, making calculations of planet and star positions from her home.

By the 1880s, more women were being hired as computers to do calculations and make measurements of photographic plates in observatories. A male graduate student of mine once quipped that American astronomy became preeminent over European astronomy because of two discoveries: Hale discovered money and Pickering discovered women.

George Ellery Hale, an eminent astronomer and organizational genius, learned how to raise money to build large, powerful telescopes by going to wealthy friends and others interested in revolutionizing American astronomy. Hale built the 40-inch refractor at Yerkes Observatory in Wisconsin, the 60- and 100-inch telescopes at Mt. Wilson in California, and the 200-inch telescope on Palomar Mountain in California. Though Hale's efforts helped put Americans at the forefront of astronomy, Hale harked back to the nineteenth century in his attitudes toward women. He and other astronomers dubbed the living quarters on Mt. Wilson (and later Palomar) The Monastery and banned women from using the telescopes—a restriction not lifted until the mid-1960s.

Edward C. Pickering, as director of the Harvard College Observatory from 1877 to 1919, responded to the competitive forces in astronomy by combining observational astronomy and physics into a new technology— the field of astrophysics. Photographing the heavens each clear evening, astronomers used spectroscopy—examining the constituent wavelengths of a star's light through a prism attached to a telescope—to distinguish between different types of stars. Pickering needed helpers to search the thousands of photographic plates his equipment was generating and to carry out long, detailed calculations to determine the positions and other information about those heavenly bodies recorded on the plates. Planning and directing the science was a man's job; tedious detail work was considered suitable work for women amateurs. While his style of doing astronomy opened the door for employing women, Pickering's attitudes were nonetheless financially motivated. He learned that the women he hired were "capable of doing as much good routine work as astronomers who would receive much larger salaries. Three or four times as many assistants can thus be employed," he reported in Harvard College Observatory's annual report of 1898, "and the work done correspondingly increased for a given expenditure."

Eminent women too shared the view that women were less suited for scientific tasks involving creative thinking. In 1893 physician Mary Putnam Jacobi sent a paper to the World's Congress of Representative Women held in Bogota, Colombia: "Modern science," she said, requires "a great number of assistants to perform manipulations involving much labor and time, requiring intelligence and great accuracy, but not necessitating original mental power.... This is a most useful and important field of work for women."

Of all the observatories hiring women, Harvard College Observatory hired the greatest number—a total of 45 during Pickering's years as director. Along with Pickering's new approach to astronomy, the establishment of the Henry Draper Memorial also contributed directly to this surge in jobs for women.

Henry Draper, a wealthy New York doctor and amateur astronomer, took the first photograph of the spectrum of a star in 1872. Spectral lines in stars—a series of dark lines appearing across a continuous band of color that corresponds to the radiation emitted by a star—had been observed through spectroscopes attached to telescopes since the early 1800s. Later, stars were classified into several types according to these spectral lines, which indicate the star's various chemical elements. When Draper started photographing stellar spectra using a spectrograph attached to the telescope, he could make a detailed photograph of the spectrum of a single star. With that innovation, the possibilities for studying the stars took a giant leap.

Historians have dubbed the women Pickering hired to perform such meticulous study "Pickering's harem." Modifying Draper's technique, they produced telescopic images of many stars, each spread out to form a spectrum, on a single photographic plate. Using a magnifying glass, they studied the spectrum of each star in order to classify it. They recorded their observations, identified other heavenly bodies photographed with the stars, and checked the results with charts. Working with incredible patience and unflagging industry, they were observers, computers, and discoverers. Some became full-fledged mathematical astronomers, computing orbits of planets and asteroids. Some compiled star catalogs, devising systems to estimate stellar brightnesses. Some, like Williamina Fleming, were put in charge of managing the staff and hiring other women assistants.

An entry from Williamina Fleming's diary, dated March 12, 1900, tells something of her attitudes toward Pickering's policies on promotions and raises:

> "During the morning's work on correspondence etc. I had some conversation with the Director regarding women's salaries. He seems to think that no work is too much or too hard for me, no matter what the responsibility or how long the hours. But let me raise the question of salary and I am immediately told that I receive an excellent salary as women's salaries stand.... Sometimes I feel tempted to give up and let him try some one else, or some of the men to do my work, in order to have him find out what he is getting for $1,500 a year from me, compared with $2,500 from some of the other assistants. Does he ever think that I have a ... family to take

care of as well as the men? But I suppose a woman has no claim to such comforts. And this is considered an enlightened age!"

Many of the women working at the Harvard observatory were outstanding. Annie Jump Cannon established the system with which she classified the spectra of more than 350,000 stars. Cannon would examine the plate with a magnifying lens, mentally classify the star into a number of alphabetical categories depending on the pattern of lines she saw, and call out her identifications to an assistant, who would write them down. She learned to identify the line patterns almost instantaneously, at a rate of more than three stars a minute. Arranging the spectral types of stars in order of decreasing temperature, she originated an alphabetical sequence that was ultimately rearranged into the hottest to coolest sequence O, B, A, F, G, K, M—the Oh Be A Fine Girl Kiss Me sequence that every beginning astronomy student today must learn. The results of her classifications are published in a work named, ironically, *The Henry Draper Catalogue*. This compilation laid the groundwork for modern stellar spectroscopy.

In 1925, Cannon received, among other honors, the first honorary degree Oxford University ever bestowed on a woman. But through four decades of work at the observatory, she received no academic recognition from Harvard. Not until 1938, shortly before her death, was she made a professor of astronomy. As early as 1911, a visiting committee of the observatory reported: "It is an anomaly that, though she is recognized the world over as the greatest living expert in this line of work ... she holds no official position in the university."

Henrietta Swan Leavitt joined the observatory staff permanently in 1902. In 1910 she made perhaps the greatest discovery of the Harvard women of this era. She identified the Cepheids—stars in the Magellanic Clouds whose brightnesses vary. In so doing, she discovered that the period of a star's variability was related to the star's intrinsic brightness. The longer the cycle from faint to bright to faint, the truly brighter the star. This discovery evolved into the most fundamental method of calculating distances in the universe: by observing the period of variability of stars in other galaxies and thus obtaining their true brightness to compare with their apparent brightness. This made it possible for Edwin Hubble to later demonstrate that our galaxy is only one of billions in the universe. Obtaining distances to other galaxies by this method will be one of the prime tasks for the Hubble Space Telescope.

However, Leavitt was not permitted to pursue her discovery; her job was to identify and catalog the variables. Pickering also assigned her the difficult job of comparing color indices and magnitudes on plates from different telescopes. According to Cecilia Payne-Gaposchkin, another of the eminent women astronomers who came later to the observatory, this was a "harsh decision, which probably set back the study of variable stars for several decades, and condemned a brilliant woman to uncongenial work." Leavitt died at a young age, before Professor Mittag-Leffler of the Swedish Academy of Sciences would be able to nominate her for the Nobel Prize he thought she deserved.

By 1920, American women could study science, though generally only in women's colleges; a few could get graduate degrees; and a dozen or so women had earned Ph.D.'s in astronomy. But the belief persisted that the role of women in doing science was different from the role of men. In a graduation address delivered to the 1921 class of Bryn Mawr College, Simon Flexner, Director of Laboratories at the Rockefeller Institute, discussed "The Scientific Career for Women." He distinguished discoveries based on "genius" or "imaginative insight"—and here the scientists he mentioned were men—from the predictable discovery demanding "knowledge, often deep and precise, and method, but not the highest talent." Here his example was Madame Curie.

Cecilia Payne-Gaposchkin received in 1925 the first Ph.D. in astronomy Harvard granted. Her thesis on stellar atmospheres was described by Otto Struve, an eminent astronomer at Yerkes Observatory at the time, as "undoubtedly the most brilliant Ph.D. thesis ever written in astronomy." She chose to remain at Harvard, since few other positions were available to her. But her career there was orchestrated by the observatory directors. She virtually never obtained the freedom to choose her own research directions, and her achievements were less remarkable than they might have been. For most of her professional career she remained untenured. Like Cannon, she was made a professor of astronomy and granted tenure at the end of her career.

Late in her life and early in my career, I attended an international astronomy meeting at the National Academy of Sciences at which she was present, and one evening found myself helping her fix her zipper in the ladies' room. Impulsively I took the opportunity to ask her many questions concerning her experience as a woman in a scientific field dominated by men. Oh, no, she replied to each of my questions, being a woman had made no difference.

But the next evening she sought me out as we were socializing in the Great Hall before the banquet. "You know those questions you asked me last night?" she asked. "Well, I decided that I gave you all the wrong answers." Then she proceeded to describe many of the difficulties that had plagued her throughout her career at Harvard. Her autobiography, *The Dyer's Hand*, published after her death, tells a tale of disappointment after disappointment, of opportunities denied. One of the most brilliant astronomers of her time, Payne-Gaposchkin was never permitted to work on astronomy's significant problems and never elected to the National Academy of Sciences.

By 1950, women astronomers with Ph.D.'s from American universities numbered about 50 in a total community of about 300. Almost all of them were employed by women's colleges; a few had access to other opportunities through a father, uncle, or brother who could sponsor them in the world of science. Almost all were single. They could look back on 100 years of American women doing astronomy and note that limited opportunities had generally restricted the contributions women had made. They could not know that as a total percentage of the astronomy community their numbers would soon begin to shrink. At the founding of the American Astronomical Society in the 1890s, the 11 female charter members constituted about 10 percent of the society. By 1985 women members numbered about 300 out of 4000—about 7%.

Since the 1950s, opportunities for women in astronomy have increased, but serious problems have not disappeared. Women whose brilliance is apparent at an early age can study at prestigious undergraduate universities, be accepted to graduate schools, accomplish important research, and obtain university or observatory positions. But, as with their male counterparts, such brilliance is rare. The remainder suffer because of their small numbers.

A student who thinks she might like to be an astronomer will often enter a department where she will be the only woman student; there will be no women on the faculty. If fortunate, she will find a sympathetic adviser and congenial colleagues with whom to study. Even so, she will be treated differently from male students. One faculty member may proclaim openly that he doesn't want a woman to work with him. Her work will be scrutinized with a care that most of her male classmates will be lucky enough to escape. She will stand out in everything that she does. And if she persists and obtains a degree, her adviser may well sit her down and suggest that she not set her sights too high in seeking a postdoctoral position.

This kind of gatekeeping also serves to limit opportunities. The letters of recommendation that her adviser writes will not be discriminatory but may be subtly different and tentative. If she is married, she may not receive job offers: "We thought her husband would not want to move" is the usual excuse. And when she goes to a meeting, she is likely to be the only woman attending.

Permanent jobs in astronomy are scarce and hard to get for young men and women alike. Affirmative action seems to have made few inroads in

the filling of academic positions. It is common for an astronomy department to receive 100 or more applications for a job; usually no more than
one or two of the candidates are women.

Women constitute only a tiny fraction of tenured professors of astronomy.
Many important astronomy departments, such as Harvard's, and the Mt.
Wilson and Las Campanas Observatories of my own Carnegie Institution
of Washington, have no women on their permanent staffs. I think this is in
part because the field of astronomy is still so dominated by a male establishment. A single member of a department search committee who is re-
luctant to add a woman to his staff can have an enormous influence for
many years. Cases have occurred in which an application list of many has
been carefully narrowed down to three: two men and one woman, in that
order. Following job offers to the top two, who decline the offer, the decision is then made to reopen the competition rather than offer the job to the
third. Rarely does this happen when the top three candidates are male.
Unfortunately, as the job market becomes even tighter, it is unlikely that
the number of women in tenured academic positions will increase.

The saddest part, of course, is that only about one-fifth of the women
who enter college intend to study science. Lack of support and encouragement at an early age has by then taken its toll. A young woman who enters
graduate school to study science is a rare creature indeed, to be encouraged and supported. But instead, the colleges are often a part of the problem rather than part of the solution.

In spite of these difficulties, women are becoming astronomers—and
successful ones. They are asking important, imaginative questions about
the universe and getting answers no less often than their male colleagues.
Only for the past 20 years or so have they been permitted to apply for
telescope time on all telescopes—time being allotted on the basis of the
excellence of the proposal. Now about one-third of the telescope time of
the national facilities, which include Kitt Peak Observatory outside of
Tucson, Arizona, and Cerro Tololo Observatory in Chile, is assigned to
women.

A cable that was sent to me in 1978 is a testament to that. "Dear Madame," it reads, "You might appreciate hearing that four women astronomers are observing on Cerro Tololo tonight, on the four largest telescopes!
We are M. H. Ulrich, M. T. Ruiz, P. Lugger, and L. Schweizer." I hope the
sky was very clear that night.

Opening the Doors

1990

When Judy Franz called me in Cambridge, England, where I was spending the summer, to invite Judy Young and me to give a joint dinner address at this meeting on Recruitment and Retention of Women in Physics, it was easy to accept. But later, when I reflected upon a talk that would both entertain and perhaps inform, I wondered "Why me"? Is it because I have been an activist during all of my professional life when it relates to advancing the cause of women in science? Is it because I can look back upon many joyous years of combining astronomy with a family devoted to science? Is it because our daughter is also an astronomer [while her father is a physicist and her brothers are geologists (2) and a mathematician]? But for any reason, it is fun to share the platform with a daughter, something Judy and I have not done many times. Probably the first time was when she and her older brother came to my Ph.D. commencement. Years later I chaired a meeting at which she spoke, and a few years ago we were both on a panel at Vassar college, and that was pleasant. For tonight's talk, she and I made a few feeble attempts to do something in concert, but we quickly gave that up. So I will start, and she will follow, each unaware of what the other is going to say.

There is a children's story called *The Paper Bag Princess* written and illustrated by R. N. Munsch and M. Martchenko. In this story, the princess is about to marry the prince when a dragon appears and carries off the prince. The princess follows a road littered with charred trees and drying bones to find the dragon, outsmart it, and finally rescue the prince. By that time, she is disheveled and dressed only in a paper bag. When the prince sees her, he says "You look a mess, your hair is a mess, your face is dirty, and you are dressed in a paper bag. When you are cleaned up and look like a princess, I will marry you." To which the princess replies, "Ronald, you

look like a prince but you are really a bum. I won't marry you after all."
She would have made a great scientist.

I live and work with three basic assumptions: (1) There is no problem in
science that can be solved by a man that cannot be solved by a woman. (2)
Worldwide, half of all brains are in women. (3) We all need permission to
do science, but, for reasons that are deeply ingrained in history, this permission is more often given to men than to women.

Once, in preparing for a talk, I asked women scientists for the most
outrageous statement made to them because of their sex and their profession. Here are a few that I can politely repeat:

"In my day we didn't have any contaminants," stated a 90-year-old scientist who returned to his former laboratory and found a young woman
crystallographer working there. Incidentally, this young woman is an ex-
perimentalist working at the interface of geology and biology. About a
year ago she had a baby. While nursing the child, she became curious
about the relation of food intake to bone and protein production in infants,
so she started snipping and examining the baby's fingernails; ultimately
she was able to differentiate breast-fed from bottle-fed babies, and thus
invented a new field of research. Her work received immediate publicity,
and is of importance to those seeking to learn from fossil records about the
diet of early animals and people.

"You can sit on my lap, honey," was said at a crowded meeting.

"I'm not going to sea with a girl on my boat" said the sailor to the
oceanographer, to which the leader of the expedition replied "OK sailor,
stay home."

"Go and find something else to study," said the department chairman to
the young woman entering graduate school.

"Why don't you just go off and get married," was the counsel the young
advisor gave to the woman student who came to him with a problem.

"That's all right, little girls are never good in arithmetic," said the first-
grade teacher to the young mother.

"You do science because we let you"; where "you" refers to women and
"we" to the establishment. The speaker was later president of an American
scientific society.

"You should do OK as long as you stay away from science," proclaimed
my macho high-school physics teacher when I told him I had gotten a
scholarship to college. He was unaware of my serious interest in becoming an astronomer; his physics classes and laboratories were a boys' club
in which the few girls present were politely ignored. Also while in high
school, I wanted to take a class in mechanical drawing, a class that was
given in the boys' wing of the school, near the wood shop and metal shop

and boys' gym. I was smart enough to talk a girl friend into joining me, because the prospect of going alone scared me. We completed the course, with great enjoyment and some ability, and I still make most of my own figures for my papers.

And then there are the little "playlets" we are all unwilling participants in: the eminent scientist who is assumed to be the wife when she arrives at the registration desk with a colleague; the woman in a group whose hand is left dangling as the new arrival shakes everyone's hand but hers; the advanced professional who is offered a job at a fraction of her current salary.

It is hard to know if we should laugh or cry at these tales. As women scientists we ride a rollercoaster of highs and lows, as progress and failure are reported, often in the same breath. But let me try to put up some sign posts for where we are going.

What are we doing right? Women are studying science in larger numbers, meetings have more women in the audience, some professional societies have made wonderfully successful efforts to insure that women are on committees, on ballots, on platforms, and behind podiums. At the United States national observatories, where access to telescopes is on the basis of peer-reviewed research proposals, women are obtaining telescope time in excess of their numbers in the field. So I am cautiously optimistic. But women are still such a small fraction of the scientific endeavor that they suffer all the social ills and disadvantages of other minorities.

What are we (the science establishment) doing wrong? I think our greatest failure is in not getting the fun and excitement of doing science across to the young; too many think that science is not for them. We have to show young people that science is not drudge work in a dark, lonely laboratory. We too seldom stress that science requires imagination, creativity, and ardor.

We are failing by not giving little girls role models early on. By age two or three, little girls know that men are doctors and women are nurses; it takes effort on the part of parents to show them otherwise. Television is the worst offender I know in this regard, and I shudder at TV commercials in which grown women discuss the color of their sinks, their floors, their wash. I long for commercials showing men doing the ridiculous and silly things that women routinely are shown doing; meanwhile the set remains off.

We are failing by not nurturing every girl who enters college wanting to be a scientist. Such a student has already done something unusual by resisting peer pressures as well as society's pressures to conform to the TV role. Yet every woman who enters college interested, prepared, and intending to become a scientist, and then turns to another field, reinforces the view that colleges are often part of the problem, rather than part of the solution.

We are failing by not recognizing the value for science of talents other than problem-solving talents. In astronomy, as I suspect in other sciences, among the most important criteria for achievement are creativity in devising programs for study, ability to see connections, a good memory, perseverance, and lots of energy.

We are failing by not giving young women the self-confidence to think that they can be scientists. Achievement is tied to expectations. We must offer women a warm welcome in science, and the possibility of success.

My advice to women students: don't quit. Muddle through. Get your "union card" (Ph.D) if you want to do research. Don't think you can't succeed if you're not first in your class, or even in the middle, or even below that. Academics admit to being notoriously poor in predicting which of their students will succeed in science. You will increase your confidence as you go along.

Don't think you have to be like all other scientists. You can be yourself. There are enormously diverse styles of doing science, and the variety will increase as science becomes more egalitarian. You can help make the change. Recent discussions on science and gender have made this issue prominent. It is style and imagination that separates scientists and their work one from the other, more than ability.

Learn the history of women in science, and tell it to your colleagues. Maria Mitchell's discovery of a comet in 1847 should be a part of our scientific cultural heritage, just as is Franklin's kite. Help to make it so. Margaret Rossiter's *Women Scientists in America: Struggles and Strategies to 1940* is a good place to start.

As far as you can, protest every all-male meeting, every all-male committee, every all-male department, every all-male platform and program at meetings. Although many may disagree with this position, I routinely advise women undergraduates not to enter graduate departments that have no women faculty, and not to enter departments where they will be the only woman student: it's too difficult. Equally important, I urge the students to tell the college the reason for their action. It is one of the very few weapons for change available to young women students.

Finally, to prove that things are really getting better, I quote from a leaflet published by the Astronomical Society of the Pacific, in the year 1944, entitled *Astronomy—The Distaff Side*. It starts out, "Among many firm convictions rudely shattered by World War II is the one that women are unfitted for jobs of a highly technical nature." I interrupt to note that women, in moderately large numbers, had been working at the Harvard

College Observatory (at one-fourth the pay of men) since before 1900, doing work of a highly technical nature. They had been doing similar work other places as well; hence, the need to know our history. The leaflet continues, "'I have gotten accustomed to the idea of 1000-plane bombing raids, radar, and rocket guns,' a friend confided recently, 'but the sight of a girl reading logs off a slide rule still has me baffled.'" The writer, a staff member at the Mt. Wilson Observatory, then attributes the lack of women observers at telescopes to the strain on the observer's powers of endurance. He could of course not forsee that it would not be until 1965, 21 years later, that the Mt. Wilson and Palomar Observatories would finally permit women to use their telescopes. He ends by asking what happens to all the women who study astronomy, and concludes (italics his) *Most of them become astronomers' wives instead of astronomers!*" For any who wish to be an astronomer rather than marry one, and who are willing to work for it, I hope your lives will be filled with many options.

Sofia Kovalevskaia: Scientist, Writer, Revolutionary[a]

1985

Wen we studied calculus in high school we learned about Leibniz, Newton, and Descartes; we learned that the only mathematical figure named after a woman is the Witch of Agnesi. The mention of the witch always brought hearty laughter. We laughed too back then, without the overwhelming sense of sadness that we feel today following such episodes. Today, we would stand up and discuss the virtually insurmountable problems faced, until recently, by any woman who attempted to enter the profession of science or mathematics.

If you doubt this statement, read the biography of Sofia Kovalevskaia (1850–1891) by Ann Hibner Koblitz [3], and ask yourself whether you could have been a mathematician given the limited opportunities available to Kovalevskaia. Her talent for mathematics appeared early. In her autobiography, *A Russian Childhood* [4], she relates a curious tale. When in 1858 her family moved back to the family estate, wallpaper carried from St. Petersburg was insufficient to cover her room. Instead, the room was papered with copies of her father's calculus notes. As a child, she attempted to follow the order of the pages, and to puzzle out the strange symbols.

By age six she had taught herself to read, for lessons were not forthcoming. At fourteen, she developed trigonometry (by substituting a chord for a sine) so that she could understand a physics text. At eighteen, living in Russia where women were not admitted to the universities, and in a society which would not permit a young unmarried woman to travel, she arranged a marriage for herself. Now she could go abroad to enter a university and could act as a chaperone to her older sister and various talented

---

[a] With Linda L. Stryker.

Figure 1. *Sofia Kovalevskaia in her mid-twenties (about 1875).*

unmarried friends so that they too could have educational and professional opportunities. When mathematicians at the University of Vienna would not accept her as a student, she traveled to Heidelberg where university authorities ultimately gave her permission to attend classes if the professors agreed.

Following three semesters of study and upon the recommendation of her Heidelberg mathematics professors, Königsberger and DuBois-Reymond, Kovalevskaia (Fig. 1) went in 1870 to Berlin for graduate study with Weierstrass. She easily solved the problems he gave her, and Weierstrass was later to write that her solutions indicated the "gift of intuitive genius to a degree he had seldom found among even his older and more developed students" [3, page 99]. Although the University of Berlin also refused to accept her as a student because of her sex, Weierstrass agreed to tutor her; lessons continued twice a week for almost four years. Weierstrass was to remain her friend, her mentor, her entry into the European mathematical community, and her correspondent throughout her life.

After the years of intense work in Berlin, Weierstrass convinced University of Göttingen officials that Kovalevskaia merited a doctorate in mathematics on the basis of three papers: the proof of a theorem in partial differential equations now referred to as the Cauchy–Kovalevskaia theorem; a study of the reduction of abelian integrals to simpler elliptic integrals; and a study of the classical problem of the shape of Saturn's rings. Kovalevskaia thus became the first woman since the Renaissance to obtain a Ph.D. in mathematics. The proof of the Cauchy–Kovalevskaia theorem is now considered one of her two most important works. Hermite called her formulation "the last word" on the subject, and said that the paper was now regarded as "the point of departure for all future research in partial differential equations" [3, page 122].

But even this achievement did not secure her the prestigious academic post she expected following her return to Russia. When academic positions were not forthcoming, she and her husband turned to complex financial schemes coupled with a social life which encompassed literature, art and the theatre. By 1878, after several such years, Sofia resumed her correspondence with Weierstrass, and became a vagabond mathematician, moving first to Moscow, then Berlin, then Paris, never with a professional position but always working at mathematics and interacting with mathematicians in the community. Of the years following Kovalevskaia's husband's death, Koblitz writes, "As a widow, Sofia would encounter fewer obstacles to her mathematical career than as a single or married woman" [3, page 174].

Finally Kovalevskaia was offered a position at Stockholm University through the efforts of the mathematician Gösta Mittag-Leffler (1846–1927), also a student of Weierstrass. Although she had to learn Swedish in order to lecture, she soon triumphed as a mathematician, a writer, a famous "woman professor," who was ultimately awarded the prestigious Prix Bordin of the French Academy of Sciences for her study of the rotation of a solid body about a fixed point. This subject is now referred to as Kovalevskaia's problem. Yet throughout these later years, until her early death in 1891, she was restless, often exhausted by her continual traveling and searching for what never came—a position at a Russian university.

To help us understand the world in which she moved, and the social attitudes surrounding this remarkable woman, Ann Hibner Koblitz has woven the story of Kovalevskaia's life into that of the political history and mores of her time. This lively story encompasses the Nihilist movement, the Paris Commune (where Sofia and her husband arrived in 1871, having crossed the Seine at night under fire), the mathematics and mathematicians of her day, and above all the role that gender played in her efforts to be a productive member of the mathematical community. It is from this latter perspective that Koblitz succeeds best; it is from this view that we would like to discuss the parallel efforts of her American contemporaries to gain entry into academic communities.

On the dust jacket of the book is a quote from Mark Kac: "Seldom does a biography of a mathematician pack so much drama and offer so much exciting reading. We must all be grateful to Ann Koblitz for giving us this book about the first great lady of Mathematics." It is easy to find reasons why great women mathematicians did not develop prior to Kovalevskaia's era. It takes both a natural ability and an entry into the contemporary community of scientists in order to make significant contributions to mathematics or other science. Rarely does the isolated, solitary worker produce a body of work which entitles him or her to be called great. Because the academic communities of the 1850s and 1860s permitted Kovalevskaia and a few others to fight their way in, some of these women were able to contribute to the scientific endeavor to a degree previously not possible.

Maria Gaetana Agnesi (1718–1799) was one of those earlier talented women who lacked such opportunities. The eldest child of a professor of mathematics at the University of Milan, she raised and tutored her 20 siblings following her mother's death. By 1738, at the age of twenty, she had embarked upon her most important work in mathematics, "Le Instituzioni Analitiche," conceived as a textbook on differential and integral calculus for her younger brothers. In it, she brought together the methods of Newton on fluxions, Leibniz on differentials, as well as the work of Fermat, Descartes, and Euler, among others. In this work Agnesi discussed the versed sine curve, originally described by Fermat, which has made her name familiar to all students. This curve, $xy^2 = a^2(a - x)$, had come to be called a *versiera*, a word derived from the Latin *vertere*, "to turn." But *versiera* also happened to be a shortened form for the Italian word *avversiera*, or "wife of the devil." In 1801, when Agnesi's book was still important enough for English translations to be made, John Colson took a turn for the worse and rendered *versiera* as "witch"; the curve continues to be known as the Witch of Agnesi [5].

At age twenty, Agnesi requested permission of her father to enter a convent to spend her life studying and working with the poor. Her request denied, Agnesi devoted the remainder of her life to helping the poor. We cannot know what mathematics she might have produced, had she been encouraged; her mathematical talent was forced to remain undeveloped. Every such loss diminishes our science. And we will never know whether it was with agony or ecstasy that Agnesi received communications such as that from the French Academy of Sciences stating that she would have been made a member of this learned body had it not been against the constitution to admit women.

Fortunately for women interested in obtaining advanced education, the attitudes of society in the 1850s offered more hope than those of the 1750s. As Kovalevskaia was aware, she was one important participant in a world-wide effort to have women accepted as students. This movement was especially active at both the undergraduate and graduate levels in the United States, where public education for women had begun to flourish during the 1820s. Women's academies, like their counterparts for men, placed strong emphasis on science; at least six women's schools had astronomi-cal observatories before 1860 [7]. Thus, even though jobs for women in science were virtually nonexistent, the study of science became an appropriate occupation for middle-class women.

In Europe, Kovalevskaia was hailed as a symbol of woman's emergence into the public world of science. Her equivalent in the United States was the astronomer Maria Mitchell [2]. Born in Nantucket in 1818 and employed as librarian of the Nantucket Athenaeum during the 1840s and 1850s, Mitchell had access to advanced astronomical texts and the leisure to study them and absorb the detailed mathematics dealing with the motions of the planets and stars. Evenings, she and her father were on the roof of their home scouring the skies with a telescope; she became a skilled observer. On October 1, 1847, while her parents were entertaining guests, Mitchell saw a comet through the telescope, and promptly announced her find to her father. Mr. Mitchell immediately posted a note to Prof. Bond, Director of the Harvard Observatory. In 1831, the King of Denmark had made an offer of a gold medal to the next discoverer of a telescopic comet. Mitchell's discovery was adjudged to be the first and the medal was hers. She was thus following in the footsteps of many earlier astronomer comet-finders including Edmund Halley and Caroline Herschel, discoverer of eight comets.

Maria Mitchell was not without other honors and recognition. Joseph Henry, in his first Annual Report of the Smithsonian Institution in 1848, announced the "account of a new comet, the discovery of which ... is one of the finest additions to science ... ever made in this country." However, the "American lady" who made the discovery was not identified by name [1]. When Vassar Female College opened in 1865, Maria Mitchell was its first Professor of Astronomy and Director of the Observatory, even though she lacked a college degree. Mitchell was the first woman elected to the American Academy of Arts and Sciences (95 years would elapse before the next woman was so honored), and was an active member of the American Association for the Advancement of Science.

During the 1860s and 1870s, Mitchell joined with similarly distinguished women scientists to educate in the women's colleges the first significant numbers of female scientists produced in the U.S. academic community. The next goal of these young women, entry into graduate school, was virtually unattainable. Many of their professors had themselves been denied permission to attend graduate courses. These professors and students now carried the battle to two fronts: to distinguish themselves by their scientific research, and to make it easier for future generations to obtain the graduate education they were denied. Their successes and failures are impressively documented by Margaret Rossiter [6] in *Women Scientists in America: Struggles and Strategies to 1940*, a story which repeatedly illustrates the universality of Kovalevskaia's circumstances. Especially interesting are the numerous efforts made by American women students to enter German graduate schools; graduate schools which ultimately accepted them because they posed no threat to the German academic job market following their graduation and return to the United States. But upon their return, these women used their acceptance by foreign universities as a wedge to get American graduate schools to accept women students.

The first U.S. doctorate awarded to a woman went to Helen Magill at Boston University in 1877. By 1889, ten colleges had awarded 25 Ph.D. degrees to women; six were in science or mathematics. We know little about the circumstances leading to these first advanced degrees from Boston University, Syracuse, and Wooster College. It was only in the early 1890s that Brown, Chicago, Columbia, Pennsylvania, Stanford and Yale opened their graduate schools to women. By 1900, nine doctorates in mathematics had been awarded to women.

The University of Berlin, which in 1870 had refused Kovalevskaia the privilege of attending lectures, 20 years later permitted Ruth Gentry (Michigan, 1890) to attend graduate lectures, but not to be a candidate for a degree. Gentry later became Associate Professor of mathematics at Vassar and in 1894 was one of the founding members of the American Mathematical Society.

By 1891, the year of Kovalevskaia's death, moderate success could be claimed; qualified women were permitted to enter some U.S. and foreign graduate schools. However, the dual battlefronts of education and job opportunities did not disappear. Princeton University did not admit women to graduate school in mathematics until the late 1960s. It would be interesting to learn how the mathematics community would have been affected had Princeton University graduate school admitted *no* mathematics students during these years.

By every measure, Kovalevskaia was a remarkable woman, a talented mathematician, and a leader in the movement to have women accepted as students and professors in academia. Had Kovalevskaia lived a few decades earlier, even her great talent would almost certainly not have resulted in a Ph.D. degree, nor in a university position. Had she lived a few decades later, she might have had fewer personal battles, and her contributions to mathematics might have been correspondingly greater. Kovalevskaia is the first great woman of mathematics because her genius coincided with an academic climate which permitted her entry. How many other gifted scientists and mathematicians have we lost through the ages because theirs was the gender which was denied permission to do science?

Letter from Caroline Herschel (1750–1848)
by Siv Cedering (in which a contemporary poet imagines herself the astronomer Caroline Herschel)

William is away, and I am minding
the heavens. I have discovered
eight new comets and three nebulae
never before seen by man,
and I am preparing an Index to
Flamsteed's observations, together with
a catalogue of 560 stars omitted from
the British Catalogue, plus a list of errata
in that publication. William says

I have a way with numbers, so I handle
the necessary reductions and
calculations. I also plan
every night's observation
schedule, for he says my intuition
helps me turn the telescope to discover
star cluster after star cluster.

I have helped him polish the mirrors
and lenses of our new telescope. It is
the largest in existence. Can you imagine
the thrill of turning it to some new
corner of the heavens to see
something never before seen
from earth? I actually like

that he is busy with the Royal Society
and his club, for when I finish my other work
I can spend all night sweeping
the heavens.

"Letter from Caroline Herschel (1750–1848)" is from *Letters from the Floating World*, by Siv Cedering, © 1984. Reprinted by permission of the University of Pittsburgh Press.

Sometimes when I am alone
in the dark, and the universe reveals
yet another secret, I say the names
of my long lost sisters, forgotten
in the books that record
our science—

> Aglaonice of Thessaly,
> Hyptia,
> Hildegard,
> Catherina Hevelius,
> Maria Agnesi

—as if the stars themselves could

remember. Did you know that Hildegard
proposed a heliocentric universe
300 years before Copernicus? that she
wrote of universal gravitation 500 years
before Newton? But who would listen
to her? She was just a nun, a woman.
What is our age, if that age was dark?

As for my name, it will also be
forgotten, but I am not accused
of being a sorceress, like Aglaonice,
and the Christians do not threaten to
drag me to church, to murder me, like they did
Hyptia of Alexandria, the eloquent, young
woman who devised the instruments
used to accurately measure the position
and motion of
heavenly bodies.

However long we live, life is short, so I
work. And however important man becomes,
he is nothing compared to the stars.
There are secrets, dear sister, and it is
for us to reveal them. Your name, like mine,
is a song. Write soon,

                                      Caroline

References

1. Bessie Zaban Jones and Lyle Gifford Boyd, *The Harvard College Observatory* (Harvard University Press, Cambridge, Mass., 1971).
2. Phebe Mitchell Kendall, *Maria Mitchell* (Lee and Shepard, Boston, 1896).
3. Ann Hibner Koblitz, *A Convergence of Lives. Sofia Kovalevskaia: Scientist, Writer, Revolutionary* (Birkhäuser, Boston, 1983).

4. Sonya Kovalevsky, *A Russian Childhood*, edited by B. Stillman (Springer–Verlag, New York, 1978).
5. Lynn M. Osen, *Women in Mathematics* (MIT Press, Cambridge, Mass., 1974).
6. Margaret W. Rossiter, *Women Scientists in America. Struggles and Strategies to 1940* (Johns Hopkins University Press, Baltimore, 1982).
7. Deborah Jean Warner, "Woman Astronomers," *Natural History* Magazine, New York, 1979.

George Gamow

1986

In 1951, my husband and I left Cornell, he with a Ph.D. and I with an M.A. My husband went to work at the Applied Physics Laboratory of the Johns Hopkins University. There he shared an office with Ralph Alpher, who was continuing his work with George Gamow and Bob Herman on the origin of the elements; Bob Herman's office was down the hall.

For my M.A. thesis I examined the 108 galaxies with (then known) radial velocities to see if there was a large-scale rotation of the entire system of galaxies. Gamow had wondered in print if such a motion could be detected. When Alpher and Herman reported my work to Gamow, he contacted me to learn the details. (Gamow later discussed this work at a colloquium at the Applied Physics Laboratory but I was not permitted to attend, for wives were not allowed past the receptionist.) I ultimately wrote my Ph.D. thesis under Gamow's direction, although he was on the faculty of George Washington University, and I was a student at Georgetown University. For my thesis, I determined the two-point correlation function for the distribution of galaxies, using the Harvard galaxy counts. Gamow suggested the problem, for he wanted to know the scale length for the distribution of galaxies.

Gamow was childlike in his enthusiasm for puzzles, games and scientific tricks. He generally carried a deck of cards and would start a lecture by attempting to pile the deck, one-by-one, successively farther over the edge of the lecture table. From another pocket would come two small balls connected by a 20-cm string; he would start the balls rotating and watch the constraints imposed. He also boasted of a membership card in a space cadet corps, obtained by sending off coupons from cereal boxes.

Gamow could not spell; he could not do simple arithmetic. I think it would actually have been impossible for him to find the product of 7×8. But he had a mind that allowed him to understand the universe.

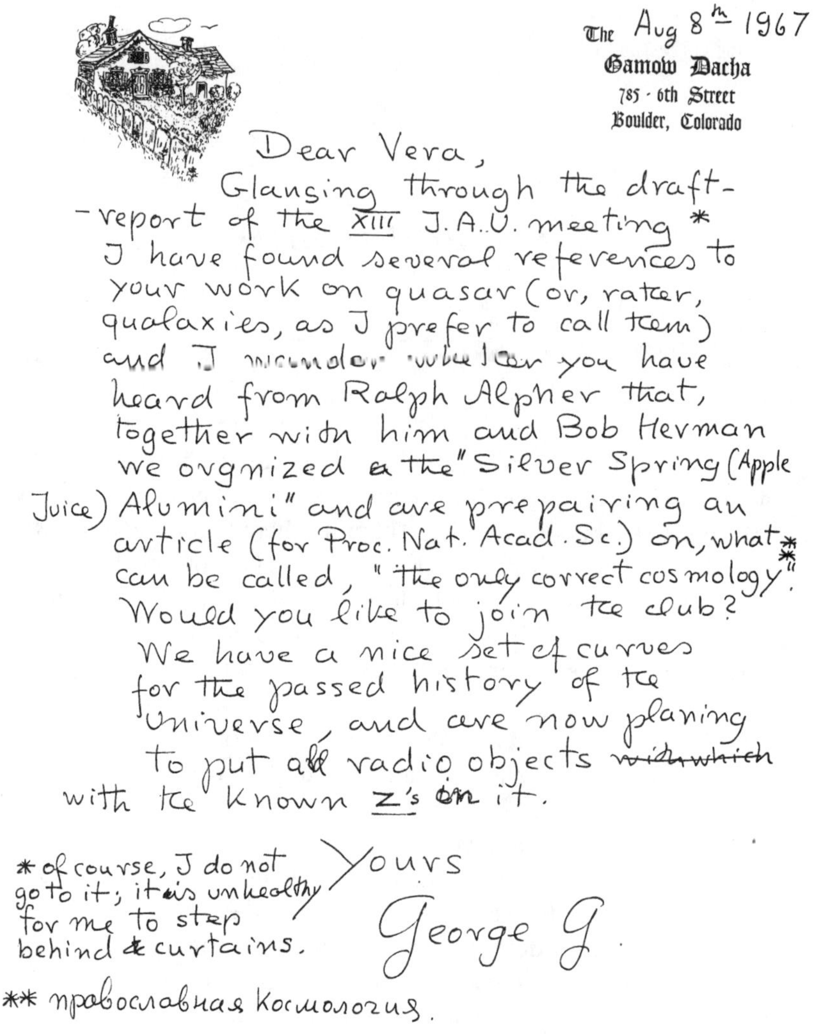

The Aug 8ᵗʰ 1967
Gamow Dacha
785 · 6th Street
Boulder, Colorado

Dear Vera,
Glansing through the draft-
-report of the XIII J.A.U. meeting *
I have found several references to
your work on quasar (or, rather,
qualaxies, as I prefer to call them)
and I wonder whether you have
heard from Ralph Alpher that,
together with him and Bob Herman
we orgnized a the "Silver Spring (Apple
Juice) Alumini" and are prepairing an
article (for Proc. Nat. Acad. Sc.) on, what **
can be called, "the only correct cosmology"
Would you like to join the club?
We have a nice set of curves
for the passed history of the
Universe, and are now planing
to put all radio objects with which
with the known z's on it.

* of course, J do not   Yours
go to it; it is unhealthy
for me to step   George G.
behind the curtains.

** православная космология.

Figure 1. *Letter from George Gamow.*

One experience stands out. In the summer of 1953, I participated in a
summer school in astrophysics held at the University of Michigan. Gamow
was one of the lecturers; Walter Baade was also a lecturer. Allan Sandage,
a new Ph.D., presented his thesis results on the color-magnitude diagrams
of globular-cluster stars. Much discussion centered on why different main
sequences were of different lengths, and why stars in different clusters
turned off the main sequence at different points.

As a very serious, conventional student, I was continually embarrased by Gamow's behavior. He fell asleep listening to lectures; woke up to ask questions regarding points already discussed; and continually asked the same questions (some of which I considered stupid). Yet by the end of the summer, Gamow had contributed in a very major way to the understanding of the effects of stellar evolution on the location of cluster stars on a color-magnitude plot. In fact, Baade credits Gamow with earlier being the first to correctly interpret color-magnitude diagrams. In a postcard to Baade, Gamow had written, "Please tell me where the lower branch of the color-magnitude diagram joins the main sequence, and I will tell you the age of your Population II stars."

One of Gamow's letters to me (Fig. 1) illustrates his whimsical, playful manner. His use of the word "qualaxies" (for quasars) in 1967 predates any clear data that identified quasars as galaxies. The "Apple Juice Alumini" is a play on the acronym APL for Applied Physics Laboratory. "The only correct cosmology" footnote identifies it as orthodox cosmology. And "not stepping behind (iron) curtains" refers to his escape from Russia with his wife in the early 1930s, via the XIII Solvay Congress. This followed an earlier abortive attempt of theirs to leave Russia via rubber rowboat across the Black Sea. His autobiography, *My World Line*, makes delightful reading.

E. Margaret Burbidge

As the next president of the AAAS, Margaret Burbidge takes her place among the many eminent astronomers who have served in this office, most recently Harlow Shapley (1947) and Walter Orr Roberts (1969). To this office she brings a distinguished career as a brilliant observational astronomer; an astrophysicist; a leader in the drive to observe from space; a spokesperson for the astronomical community, and a strong supporter of women and minorities in science. Her leadership capabilities, obvious in all phases of her career, are combined with a warm and friendly personality, and have made for her a wide circle of friends and colleagues throughout the world.

Margaret Peachey Burbidge, born in England, recalls being interested in astronomy as a youngster and reading the books of Sir James Jeans, to whom she is distantly related. She attended University College, London, and was surprised to discover that a degree in astronomy was offered. She earned a B.Sc. from University College and a Ph.D. from the University of London Observatory. Burbidge is now professor of astronomy in the Department of Physics at UCSD, and she is director of the Center for Astrophysics and Space Sciences there. She and her husband and colleague, astrophysicist Geoffrey Burbidge, have a daughter, Sarah.

Margaret Burbidge's earliest research work concerned chemical abundances in stars of various types, and culminated in the now-classic work by Burbidge, Burbidge, Fowler, and Hoyle (familiarly known as B^{2}FH) on *Synthesis of the Elements in Stars*. Willy Fowler affectionately recalls a day in 1954 during his sabbatical year at the Cavendish Laboratory when a "wonderful Charles Laughton replica" (Geoff) walked into his office following Willy's colloquium on 3-α reactions and asked, "Why not work on problems important for astrophysics?" Margaret and Geoff had spent the previous year analyzing the spectrum of the star α^2 CVn. They concluded that the overabundance of some of the elements in the atmosphere was real and that somewhere, somehow, neutrons were involved. The work required that neutrons be produced in stars, not at the origin of the universe. With Fowler they produced the very attractive hypothesis that all the chemical elements might be cooked in stars; that is, neutron capture, with the neutrons produced in stars by (α, n) reactions. This initial work required the solution of integral equations, done by Margaret on an ancient Brunsviga hand-cranked calculating machine, called by Willy the "Babbage" machine.

Ultimately, B^2FH identified eight nuclear processes: hydrogen burning, helium burning, the α-, e-, s-, r-, and p-processes, and the x-process, x for an unknown but necessary process to produce the especially difficult isotope of deuterium, and also lithium, beryllium, and boron. With this work as a basis, it is now possible in principle to reconstruct the way in which enrichment in heavy element content has proceeded in the interstellar medium, and in successive generations of stars in our galaxy.

As Fowler's return to the United States approached in 1955, he suggested that the Burbidges accompany him to Pasadena; Margaret to Mount Wilson Observatory as a Carnegie postdoctoral fellow, and Geoff as a research fellow at the Kellogg Laboratory at the California Institute of Technology. A letter to the director of the Mount Wilson Observatory elicited the response that the single toilet at Mount Wilson precluded awarding the fellowship to a woman. With their usual adaptability, Geoff, a theorist, took the Mount Wilson fellowship, and Margaret the Kellogg appointment. Not surprisingly, whenever Geoff went off to Mount Wilson to observe, Margaret "coincidently" appeared. Not until 1965 did a woman legally observe at the Hale Observatories, and not until 1979 was a woman named a Mount Wilson Carnegie Fellow. Margaret Burbidge's concerns for opportunities in science for women and minorities have a very personal basis.

A second major aspect of Margaret Burbidge's research concerns the internal dynamics and masses of galaxies. Starting in the early 1960s with observations made at the McDonald Observatory in West Texas (then jointly operated by the Universities of Chicago and Texas), she obtained spectra of spiral galaxies from which were measured the velocities of the ionized gas clouds in their nuclei and disks. She, in collaboration with Geoff and with Kevin Prendergast, ultimately deduced rotational properties and masses for 50 or so spiral galaxies. By 1970 most of the knowledge of galaxy dynamics came from studies by Burbidge and colleagues, and caused at least one scientist to thank the Burbidges in print for "several pounds of reprints" on internal motions in galaxies. Once again, Burbidge had pioneered a direction that developed into a fruitful and decisive branch of contemporary astronomy.

For the past 15 years, Burbidge has continued her observational research programs at the Lick Observatory of the University of California, in collaboration with students, fellows, and colleagues worldwide. Redshifts of quasars, absorption lines in quasars, and the distribution of quasars in the universe, all questions at the frontier of our knowledge, are among her current research interests.

With increasing frequency since the 1960s, Burbidge has acted on her belief that scientists must impart the wonder of science to the public, and must also address the problems of society and of support for science. She has taken an active leadership role on many committees for space sciences; on setting scientific priorities, and she is now a member of the Committee on Science and Public Policy of the National Academy of Sciences. She is a co-investigator on the team to build the Faint Object Spectrograph for NASA's Space Telescope.

Burbidge's achievements have been recognized with honors, prizes, and honorary degrees. She shared with her husband the Warner Prize in Astronomy in 1959; she served as president of the American Astronomical Society 1976–1978; and she was elected a Fellow of the Royal Society of London in 1964; and to membership in the American Academy of Arts and Sciences in 1969. She was elected a member of the National Academy of Sciences in 1978, the only woman astronomer so honored, and to the American Philosophical Society in 1980.

Throughout her professional career, Margaret Burbidge has met each challenge with intelligence, with originality, with dedicated hard work, and with grace. Members of the AAAS can expect an active leadership during her presidency.

The Leviathan of Lord Rosse

1993

Durring my graduate student years at Georgetown University, I stumbled across a wonderful book in the astronomy library. The book was *The Scientific Papers of William Parson, Third Earl Of Rosse 1800–1867*. I was fascinated by the details of his observations of galaxies, and by the sketches that accompanied the text. So much so that in one of my earliest papers I included a quote from his publications. Lord Rosse is a hero of mine.

At Birr Castle in Ireland in the 1840s, the Third Earl of Rosse conducted experiments in telescope construction, assisted by a carpenter, a blacksmith and his own farm laborers. After smaller successes, he ultimately built the great 72-inch telescope that came to be called "The Leviathan of Parsonstown." The telescope mirror was made of speculum, an alloy of tin and copper, and was polished with an instrument of his design with the different motions "obtained by cogwheels and leather bands." Until 1917 and the advent of the Mt. Wilson 100-inch telescope, this was the largest telescope in the world. With it, Lord Rosse and his assistants made observations of galaxies and, most notably, discovered the spiral nature of galaxies in 1845.

Rosse's publications in the *Philosophical Transactions of the Royal Society* resemble works of art with sketches of galaxies so accurate that the galaxies can be individually identified today by knowledgeable students. In addition to sketches in the notebooks, there are detailed engraved plates of the major objects. An indication of the care with which the manuscripts were assembled can be found in the final paragraph of his 1861 paper, "On The Construction of Specula of Six-Feet Aperture; and a Selection of the Observations of Nebulae Made with Them." "The engravings of the nebulae are extremely faithful: there is, however, a slight inaccuracy which it is necessary to notice, and for which we are to blame, not the engraver. Many of the principal stars are too large. The error arose in this way. The stars were inserted in common, not India ink, and, the drawings during their transmission by post becoming slightly damp, the ink made its way into the paper, the dots in some cases becoming small blots. In a few instances it was necessary to set this right to prevent misconception, and some alteration was in consequence made in the Plates; but as to the remainder, we thought it sufficient to state the fact generally, that many of the principal stars were somewhat too large."

There is also a modern story related to Lord Rosse. In an early paper on the structure of our Milky Way (see "Structure and Evolution of the Galactic System," p. 21), I concluded with a quote from his 1850 paper: "The sketches (Fig. 1) which accompany this paper are on a very small scale, but they are sufficient to convey a pretty accurate idea of the peculiarities of structure which have gradually become known to us: in many of the nebulae they are very remarkable, and seem even to indicate the presence of dynamical laws we may perhaps fancy to be almost within our grasp." I thought that this 1850 comment represented equally well the state of our understanding in 1960; that is, cautious optimism.

It was with surprise, then, that I later read the 1977 review of "Theories of Spiral Structure," by a leading expert, Alar Toomre. He quoted from

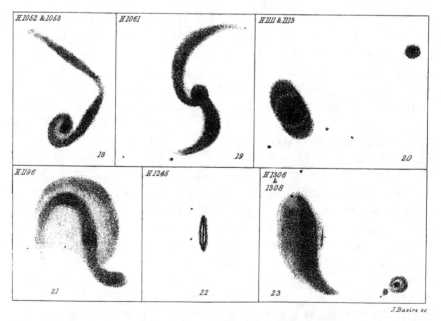

Figure 1. A collection of sketches made by Lord Rosse and his assistants, using his 72-inch telescope, and published in the Transactions of the Royal Society.

that same 1850 paper, but his statement expresses pessimism, not optimism: "Much as the discovery of these strange forms may be calculated to excite our curiosity, and to awaken an intense desire to learn something of the laws which give order to these wonderful systems, as yet, I think, we have no fair ground even for possible conjecture."

These two different quotes introduce successive paragraphs in Lord Rosse's paper! One can imagine the cause of this ambivalence as to what he knew and what he might hope to learn. His discovery of images of the many wonderous forms of galaxies were surely a puzzle to astronomers who knew neither the distances nor the compositions of these objects. Lord Rosse did not live long enough to learn of Scheiner's spectrum of the Andromeda galaxy (1899) which showed it to contain stars like the sun, or of Oepik's (1922) analysis that showed Andromeda to be a galaxy distant from our own. And how he would have liked to learn that the spiral density wave theory of Lindblad, of C. C. Lin, and especially the understanding generated by the work of Toomre and others, did indeed "indicate the presence of dynamical laws ... almost within our grasp." Lord Rosse was a tinkerer, and I imagine he would have been pleased to see computer simulations of galaxies forming and evolving; especially the wonderous forms that emerge as galaxies gravitationally interact and morphologically transform. Surely, he would have delighted in the surge of building of large telescopes that is going on today. Lord Rosse was a singular individual, and it is a pleasure tonight to contemplate our current views of the universe with his work as background.

Gérard de Vaucouleurs
and the Local Supercluster

1989

érard de Vaucouleurs did not invent the Local Supercluster all by himself. It had been hinted at years earlier by clever astronomers who had puzzled over the distribution of nearby galaxies. But it is Gérard who made us believe in it. Due almost solely to his insistence, today we know that we live on the outskirts of a flattened distribution of galaxies, whose peak density lies in the direction of Virgo. The elucidation of the structure and of the motions within our supergalactic neighborhood has occupied much of Gérard's astronomical career, and it remains one of the most active and exciting fields of current extragalactic astronomy.

In retrospect, it is difficult to reconstruct the circumstances that prevented many astronomers from accepting the concept of a supergalactic system. A paper entitled "Is the Local Supercluster a Physical Association?" was published in the *Astrophysical Journal* in 1976; it was the dying gasp of the nonbelievers. The paper had at least one valuable result. It elicited an answer from Gérard which contains the footnote, "A student of human population statistics might just as well deny the existence of New York City by assigning a large enough radius to Greenwich Village!" Today, few astronomers doubt the reality of the Virgo Supercluster as a major mass concentration in our local extragalactic environment. Much current research is devoted to attempts to understand the relation of the Virgo galaxy overdensity to similar overdense regions in Ursa Major, in Perseus–Pisces, and in Centaurus.

Since the early 1950s, Gérard has written innumerable important papers detailing the Local Supercluster. One of his early studies, "Further Evidence for a Local Super-Cluster of Galaxies: Rotation and Expansion" is especially noteworthy. In any science, early papers which grope for new truths have a vitality and forcefulness that remain evident years later. Later papers are more elegant, perhaps more correct, but the non-trivial difficulties evident at the beginning often do not go away. It is remarkable that many of the complications which Gérard discussed are still with us, and still worrysome. In many respects, this paper laid out a path of research which is still actively followed today.

Messier (1781) noted the curious fact that 13 of the 103 nebulae he catalogued were located in the constellation of Virgo. William Herschel (1784) enlarged the number of known nebulae, and concluded that the nebulae, located preferentially in a "nebulous stratum" almost perpendicular to the sidereal stratum, exhibited a strong tendency toward clustering. William's son John Herschel continued observations, ultimately culminating in the publication of two catalogs (1833, 1847) containing 3925 nebulae in both northern and southern hemispheres.

From an analysis of these data, John Herschel drew several significant conclusions:

1. The distribution of nebulae on the celestial sphere is non-uniform, with one-third of the nebulae arranged in a broad irregular patch occupying about one-eigthth of the sphere.

2. Although the central concentration is in Virgo, several other centers are also identified, one of them the Pisces region.

3. The zone of nebulae is not a zone or a band encircling the heavens, or "if such a zone can be traced out, it is with so many interruptions, and so faintly marked out through by far the greatest part of its circumference, that its existence as such can be hardly more than suspected."

Yet these hints were sufficient for Herschel (1847) ultimately to conclude that the nebulous system is distinct from the sidereal, with "Virgo being regarded as the main body of this system, and subtending at our point of view an angle of 80° or 90° ... Our distance from its center must be considerably less than its own diameter, so that our system may very well be regarded as placed somewhat beyond the borders of its denser portion, yet involved among its outlying members ..." Not a bad description of the Local Supergalaxy for the year 1847. In fact, Alexander von Humboldt made a statement even stronger than Herschel's, in his popular book on natural history, *Cosmos* (1849). "The stellar milky way, in the region of which, according to Argelander's admirable observations, the brightest stars of the firmament appear to be congregated, is almost at right angles with another milky way, composed of nebulae ... The milky way composed of nebulae does not belong to our starry stratum, but surrounds it at great distance without being physically connected with it, passing almost in the form of a large cross through the dense nebulae of Virgo, especially the northern wing, through Coma Berenicis, Ursa Major, Andromeda's girdle, and Pisces Boreales."

More than one-hundred years ago, Dreyer (1888) published *A New General Catalogue of Nebulae and Clusters* ... with positions of 7840 objects. The distribution of these objects was discussed and analyzed by numerous astronomers. Although Gérard credits Reynolds (1923) with the discovery of the Local Supercluster, it is clear that the suggestion originated as early as 1847 with J. Herschel.

However, it would be erroneous to infer that by 1950 astronomers believed that we lived in a local supercluster. When I entered graduate school at Cornell in 1948, we knew well the Harvard and Mt. Wilson galaxy observations; the Shapley–Ames (1932) catalog, the Curtis (1933) article in the *Handbuch der Astrophysik*, and Hubble's (1936) *Realm of the Nebulae* were standard extragalactic references. Other than these, I cannot now recall that I knew any of the papers mentioned above which discuss the more detailed distribution of galaxies. Radial velocities were available for about 100 galaxies; not until 1956 (Humason, Mayall and Sandage, 1956) would the number increase to over 800. Hubble's (1936) view of the local extragalactic region was overwhelmingly accepted. "The small scale distribution is irregular, but on a large scale the distribution is approximately uniform. No gradients are found. Everywhere and in all directions, the observable region is much the same." More radical ideas, some to be presented shortly by Zwicky (1957), were not abundant.

As a young graduate student interested in galactic dynamics, I wondered if the galaxies partook of a large-scale systematic motion, in addition to the Hubble expansion, a question which Gamow (1946) had raised. Specifically, I attempted to apply a first-order Oort-like analysis to the available radial velocities. My analysis was unorthodox. For each galaxy I estimated a distance from its apparent magnitude, removed the velocity due to a smooth Hubble flow, and plotted the residuals on a globe. I then searched visually and found a great circle along which there were maxima and minima of mean residual velocities. The plane identified was virtually the present Supergalactic Plane which, with the enthusiasm of youth, I named the Universal plane. Although this M.A. thesis work was presented at the 1950 AAS annual meeting, it was never published. Both the Ap.J. and the A.J. rejected it, on the grounds that the statistics were poor and the analysis open to criticism. But no one, neither my supportive M.A. thesis advisor Dr. Martha Stahr, then a recent Berkeley astronomy Ph.D., nor the physicists on my M.A. committee (Richard Feynman, Philip Morrison), nor the Journal referees, suggested (what I did not then know) that we lived in a flattened supercluster and that the existence of such a plane had been previously discussed.

I present these details to indicate the lack of general knowledge of any large scale extragalactic structure in the early 1950s. But these circumstances were also the route to several friendships. Edwin Carpenter wrote a long letter on the train returning from the AAS Haverford, Pa. meeting to Tucson, detailing his interest in and thoughts about large scale structure among galaxies. Martin Schwarzschild offered to help try to get my paper published. Ogorodnikov (1952) wrote from the USSR that he had been working along similar lines. And Gérard de Vaucouleurs started a relentless, questioning correspondence that forced me to learn more and to face problems whose troublesome existence had not previously bothered me. I don't know in what office or at what desk Gérard sat when he wrote those letters and asked those prescient questions. I do remember that I was then a Ph.D. candidate, juggling classes and a thesis with a family with two youngsters. In retrospect, I remember it like a dream in which I always owed Gérard a letter and could never find a pencil and paper.

Most of the questions Gérard was then asking, and some of their answers, show up in his very early papers. He modeled the existing anomaly in the direction of Virgo with a rotation of the local system, plus an expansion which is slowed in the denser central regions and increases outwards in regions of decreasing galaxy density. He is concerned with questions involving the distance modulus of the Virgo cluster, with the value of the Hubble constant, with velocity dispersion, velocity anisotropy, variations in the Hubble expansion in regions of high galaxy density, the mass of the Virgo complex, the Malmquist bias, luminosity functions, and the mean absolute magnitudes of galaxies of different Hubble types. Few of these questions have been resolved.

However, several important changes have occurred since 1957. The number of galaxies with known radial velocities has increased from 800 to about 20,000, and the accuracy of absolute-magnitude estimates of galaxies has increased, due to the application of the Tully–Fisher and the Faber–Jackson relations between dynamical parameters and mass. Extensive observing programs by numerous groups of astronomers, completed and ongoing, offer the prospect of a still-greater understanding of the large-scale structure and dynamics of our region of the universe. The study which Gérard initiated has grown into a major industry of extragalactic observational astronomy. A very sophisticated view of the Local Supergalaxy comes from the Tully and Fisher (1987) atlas.

One final episode. At the American Astronomical Society meeting in Mexico City, Marc Aaronson and Jeremy Mould (1978) presented a paper entitled "Application of the Infrared Magnitude/HI Velocity Width Relation to Measurement of the Distance Scale." In the talk, Aaronson described the values derived for H_0 in a few directions, making use of relatively nearby clusters. He especially noted the smaller value of H_0 derived from galaxies in the direction of the Virgo cluster. In the discussion which followed, Gérard predicted that H_0 values in other directions would be higher, but close to constant, rather than varying systematically across the sky. He based this prediction on the results from the 1958 paper described above, and on later analyses. Aaronson's paper was the final one in a session in which I was the chairperson. As Aaronson and I left the stage, Aaronson commented softly to me, "You know, he's right. We have results from a few other directions, but they're incomplete, so I didn't present them."

Like all practicing scientists, Gérard has been right, and he has also probably been wrong. For many cosmological problems of current interest, right and wrong are not yet established. Leopold Infeld, in characterizing Einstein's early cosmological works, wrote in 1947 "Indeed, it is one more instance showing how a wrong solution to a fundamental problem may be incomparably more important than a correct solution of a trivial uninteresting problem." Gérard de Vaucouleurs has dared to ask and attempted to answer hopelessly complex cosmological questions, and the field of astronomy is richer because of his daring.

Ralph A. Alpher
and Robert Herman

1993

Occasionally in the history of science, a discovery is made which is so far ahead of its time that it does not fit simply into the then-existing structure of science. Even more occasionally, such a discovery lies forgotten by the community, only to be rediscovered when needed at a later time. Such is the history of the prediction by Drs. Ralph A. Alpher and Robert Herman of the existence of a fossil remnant radiation, a remnant of the Big Bang origin of the universe.

Starting in 1948, Alpher and Herman, along with their colleague George Gamow (Fig. 1), carried out extensive studies to put physics into a nonstatic cosmological model, and to explore the consequences of such an approach. In the process, Alpher and Herman calculated the conditions that prevailed at the time in the expansion of the universe when the density of matter equaled the density of radiation. Their results permitted them to examine the time-dependence of relevant physical parameters during the evolution of the universe.

In a series of brilliant collaborative studies between 1948 and 1954, Alpher and Herman predicted the existence of the microwave background radiation throughout the universe, a remnant radiation of the Big Bang origin of the universe. In 1948 they wrote, "The temperature of the gas at the time of condensation was 600°K, and the temperature in the universe at the present time is found to be 5°K." This prediction was virtually ignored by a physics community which believed then that cosmology was capable of little more than "order-of-magnitude" calculations.

The prediction had been forgotten when in 1965 Arno Penzias and Robert W. Wilson detected the radiation serendipitously, a discovery which earned Penzias and Wilson a Nobel Prize. This discovery is widely ac-

Figure 1. A slide made by G. Gamow, showing a bottle labeled Ylem. Gamow attempted to popularize the use of the ancient Greek word meaning the initial stuff from which all matter in the universe ultimately formed. The slide was autographed for me one evening in the 1960s by Gamow, Willy Fowler (who was awarded the Nobel Prize in Physics in 1983 for his quest for the origin of the elements), Ralph Alpher, and Bob Herman.

claimed as one of the most important cosmological findings of modern times. The observation stimulated a veritable explosion of efforts to explain its origin. The correct explanation came from the work of Robert Dicke, who repeated some of the calculations of Alpher and Herman, not knowing that he was recreating a theory which had been published in the *Physical Review* 17 years earlier.

The discovery and understanding of the cosmic background radiation ranks as one of the major achievements of 20th-century science. Alpher and Herman are a part of this story, and they have earned the respect and admiration of the scientific community for their insight in developing a physical model of the evolution of the universe following the Big Bang, and especially for predicting a microwave background radiation throughout the universe.

They were scientific pioneers, ahead of their time. It is fitting that their contribution be recognized by the National Academy of Sciences with the award of the Henry Draper medal.

Desert Island Discs

J**ULIETTE WEINSTEIN:** Will the universe go on expanding, or will it collapse in a big crunch? No one knows, but for the collapse to occur there must be more matter in the universe than we can currently observe. Today's guest, Dr. Vera Rubin, contributed some of the crucial data that established the existence of this so-called dark matter. Join us now for an hour's conversation and musical choices.

J. W.: Dr. Rubin, how did you become interested in astronomy?

VERA RUBIN: I moved to Washington when I was about ten years old with my family and had a very small bedroom with a bed right under the window which faced north. When I would go to sleep at night I would look at the stars and I would watch the stars move as the Earth turned and I just got very interested in the movement that took place in the sky. There were aurora displays around that time. There were some very interesting alignments of the planets and just the experiences that I had watching the sky became very, very important to me.

J. W.: Was there any one else in your family who was involved with astronomy?

V. R.: No, my father was an engineer and he helped me build a telescope. My parents didn't object. On summer nights we drove out of the city and looked at the stars, but I knew no astronomer. I didn't know anyone who had ever wanted to be an astronomer. I really just got captivated by the sky.

J. W.: Did your friends think you were a bit eccentric?

V. R.: (chuckling) Probably, certainly yes, and as I got older, as I became a teenager I was. I had many friends, but I really survived by telling myself that I was different. It didn't seem to bother me that much.

J. W.: What did astronomy mean to you at that time? When I was a child I lived in the country and I used to look up at the sky and I was overwhelmed by the beauty of it all. Was it a very beautiful experience for you?

V. R.: It was a very beautiful experience, and it still is. Observing at a dark site where you see the sky so spectacularly that you almost can't distinguish the normal constellations is still a very very beautiful experience. I don't think my view of astronomy has changed very much and at times that surprises me. I think we tend to differentiate between amateurs and professionals and yet being a professional astronomer has really not dimmed my curiosity at all. In a sense I am an astronomer because I am curious about the sky, and I am curious about the universe and the galaxy. It is this curiosity that keeps me so intensely interested in what I am doing. And I don't think that has changed very much.

J. W.: I am looking forward to hearing about how you have progressed from a child interested in the night sky to becoming an astronomer whose contributions have been enormously significant to the field. But let's have your first musical choice today, you have chosen Marion Anderson singing "He's Got the Whole World in His Hands." What is the story behind this?

V. R.: My mother was also a Philadelphian. She went to William Penn High in Philadelphia. Her maiden name was Applebaum. The girls apparently sat alphabetically and she sat next to someone whose name was Marian Anderson. They were not really intimate friends. My mother had a remarkably beautiful singing voice and following her high school graduation, she did fairly technical things. But because she said she was called upon to sing so much, she felt that she really ought to learn to sing better, so she started taking voice lessons with someone I only know as "Bogetti." But this apparently was the same voice teacher that Marian Anderson studied with. So they became reacquainted. They attended each other's student concerts. My mother never sang professionally at all, but retained an acquaintance with Marian Anderson most of her life.

J. W.: What exactly is your position at the Carnegie Institution?

V. R.: I'm called a Staff Scientist. I work at one of the five Carnegie Institutions around the country. There are five Carnegie laboratories which Andrew Carnegie established, and I work at the one in Northwest Washington called the Department of Terrestrial Magnetism. We have a staff of 14 people; four or five of us are astronomers, the others are geophysicists, geochemists. As a staff member I am just free to do my own research. So I am supported to go to observatories, make observations, bring them back to Washington, analyze the data, and draw conclusions.

J. W.: So you do a lot of travelling?

V. R.: Yes, I do a lot of travelling.

J. W.: Do you enjoy that?

V. R.: Moderately. It is very interesting. It is also very tiring. Observing means you fly somewhere, you work through the night, you sleep very little during the day. You work very hard during the day making sure that everything is right. So after four or five nights of that you get on an airplane and come back to Washington. What I need then are just a couple of nights of good sleep.

J. W.: I can imagine. Tell me about the rotation of galaxies and your confirmation of the dark matter in the universe.

V. R.: Well, in the case of astronomers, I think it is a fair statement that probably the toughest decision you have to make is what you are going to work on. I have always been interested in galaxies. Many years ago I decided I would make a fairly systematic study of galaxies of different types in order to understand why some of them had large spiral arms, and some of them had little spiral arms. And in so doing, I started by attempting to study the way the stars orbit about the galaxy. I presume people know that when they look at the sky at night, all of the stars they see belong to our own galaxy. They are all orbiting about a very distant center. They are orbiting with velocities which differ depending upon how far they are from the center of the galaxy. They are orbiting because of the gravitational force of all of the matter between them and the center of the galaxy. So I started a program to study galaxies outside of our own, by using a telescope with a spectrograph, an instrument which spreads the light out into its component colors, much like water droplets spread the light into a rainbow. I was studying how rapidly the stars moved about the galaxy. And what I discovered was contrary to all of our expectations.

We had expected that as stars got farther and farther from the center of their galaxy, they would orbit slower and slower, this because they were farther from the mass that was attracting them. In the solar system, for example, where almost all the mass is at the sun, the planet closest to the sun, Mercury, goes faster than the others, and Pluto, which is the farthest planet, goes slowest. The Earth is actually falling to the sun, but we are always missing it because of our forward motion. That is what an orbit is. And the sun, our star, is orbiting about the center of our galaxy. It takes 200 million years at the position of our sun to orbit the galaxy once. The sun is orbiting with a speed of half a million miles per hour. These are the kinds of velocities I am measuring in other galaxies. So, the surprise was that stars very far from the centers of their galaxies were orbiting with very high velocities. And that really leads to only two possibilities: either Newton's laws don't hold, and physicists and astronomers are woefully afraid of declaring Newton's laws invalid, for they hold up in so many cases. The only other explanation is that the stars are orbiting at very high velocity because they are responding to the gravitational field of matter which we don't see. In a galaxy, the brightest part is at the center, and the brightness falls off from the center outward. So it was expected that the distribution of matter did the same thing; lots of matter in the center and rapidly decreasing amounts as you go to the outer parts. But the high velocities observed for the stars far out tell you that that is not correct. Because the amount of bright matter is decreasing rapidly with distance from the nucleus, much of this matter in the galaxy must be dark. We don't see it at all by its brightness, but we are detecting it by its gravitational attraction on the stars that we do see.

J. W.: And what is the implication of the dark matter?

V. R.: Well, the implication is that most of the matter in the universe is not radiating at any wavelength that we can observe. At least 90 percent of the matter in the universe is dark. And that is a rather daunting idea. We became astronomers thinking we were studying the universe, and now we learn that we are just studying the 5 or 10 percent that is luminous.

J. W.: I believe there are two views as to the nature of that dark matter. Is that true?

V. R.: Yes, two major camps with a variety of other views in between. One of the major camps believes that the matter could be much like the matter we know; hydrogen, helium, the atoms and molecules that we are used to observing on Earth and in the stars. Somehow, this matter has gotten itself into a configuration where it is not radiating. It could be something like black holes, it could be things that once were stars and radiated, and now have become very dark. It could be small planets which are so small, not radiating, that we can't detect them. The other camp believes dark matter to be a kind of particle that we have not yet discovered, and which does not radiate. Particle physicists are actively engaged in developing models, in developing theories, and actually building detectors so that they can attempt to reproduce some of the physical conditions in the universe. This is one of the reasons particle physicists want superconducting supercolliders.

J. W.: Is there enough dark matter to close the universe?

V. R.: That is a question we absolutely cannot answer. Theorists will tell you the universe is closed because their theories are neater that way. Observationally, even with the dark matter that is implied by the rotation curves, we have not detected enough matter to close the universe. In fact we have only seen 10 or 20 percent of the matter that would be required to close the universe. And when I say seen, I mean even by its gravitational attraction. Observational evidence to date is that we live in a low density universe which will expand forever.

J. W.: Well, I would like to return to this in a few minutes. But let's have your second desert island choice this morning. It is Hindemith Viola Sonata Opus 11 No. 4. Why are you taking this with you?

V. R.: During my life I have had lots of favorite pieces at different times. When I was in college I studied astronomy, physics, mathematics, very seriously. I also studied music. I was very interested in contemporary music. I was even very interested at one point in writing music, though I wasn't very successful at that. And through many of my college years, this Hindemith Viola Sonata was my favorite. It was very difficult to get a recording of it at that time. I don't know whether they were available, I just didn't know how to do it. And then towards the end of my college years, I met a young man who interested me very much. And among the many things he did that were very nice was that he found a phonograph record of this Hindemith Viola Sonata. He is now my husband, and this has become a favorite of both of ours.

J. W.: Your third choice is Sebelius Violin Concerto in D Minor, Opus 47. Why are you taking this with you?

V. R.: My youngest son is a violinist; a really very talented violinist. During his elementary and junior high and senior high school days in Washington, D.C., he played with the D.C. Youth Orchestra. In his senior year he won the Concerto competition playing the Sebelius Violin Concerto, and that meant he performed it with the Orchestra. Unfortunately, the evening that he performed it, I was in Chile observing. In fact, I was riding that night in the prime focus cage of the telescope; that is when you observe at the very top and you move around with the telescope. A kind of exotic way to do science. And so through the night, doing my science, I was thinking about this violin concerto that I was missing. The next day I finished observing, went down to La Serena, the little town nearby, had dinner with some friends. I told them my son had been playing the Sebelius violin concerto the night before. Without telling me, someone went to the record player, and put the record on. So sitting there in Chile, the night after Allan had played this, I heard it, and it has been a great favorite ever since I heard him practicing it.

J. W.: Dr. Rubin, you chose that because of your son Allan. He must be very talented to be able to play that.

V. R.: Well I think so, he is a talented violinist. We tried to tell him we would be very happy to have a musician in the family. We thought that with older brothers and a sister as scientists he might feel like he too had to be a scientist. But he said it was too hard to be a violinist and he could do better being a geologist just playing the violin on the side, which is what he does.

J. W.: You, in fact, have four children, who are all Ph.D.s. What on earth did you do to get four Ph.D.s?

V. R.: Well my husband is a physicist who enjoys his work as much as I enjoy mine. And I think what we did was just let the children see how much we were enjoying what we were doing. In fact, when Allan was about 10 years old, one night at dinner he said to me "Do they pay you for the work you do at Carnegie?" That made me realize he didn't know many people that had as much fun as I seemed to be having and were paid for doing it.

J. W.: Yes, people who enjoy their work are very lucky. And I believe that music is a very important part of your work.

V. R.: Music I think is important to almost all observers. Dark, cold nights, standing at the back of a telescope, which is what we did until five or ten years ago, can be very, very long. There were many nights I would take, during twelve hours, two six-hour integrations. I would take two exposures during the night, standing in total darkness, no one else around. My eye is glued to the crosswire, keeping a star at the center of the crosswire so that the telescope is accurately guiding. There really is nothing much that you can do except listen to music. And for the earliest years of my observations, even that was difficult. There were only a few programs, like American Airlines Music Till Dawn, and you would have to pick these up. It was really very hard. And then of course ultimately observatories became equipped with either tapes or record capability. And much of what we do during the night besides sitting there and keeping the telescope going is to listen to music. It means you listen to very long things, operas, long symphonies, even Gilbert and Sullivan at 4:00 in the morning to stay awake. The Pirates of Penzance for many years was sort of my stay-awake 3:00 or 4:00 a.m. music.

J. W.: Let's have your fourth piece of music. You have chosen the Mahler 5th Symphony. Why have you got this with you?

V. R.: I think at the present time, the Mahler Symphonies are among my favorite pieces. Maybe because of their length, so that they are good telescope listening, I am just very fond of them and I think of all of the symphonies, I am fondest of Mahler's Fifth.

J. W.: Can what we learn from astronomy be put to any practical use here on Earth?

V. R.: There are things we learn by studying the universe that might have very real applications. The way in which energy is produced in stars may be a way in which energy might be produced on Earth. It has been suggested that we could, at some future date, gather more of the sun's energy, more of the sun's radiation, than we do by just being in the way. We could put reflectors up. There are reasons why this might be good, there are reasons why this might be bad. But there are all kinds of practical implications, certainly in the solar system. By the time you get very far out it is hard to understand right now what the practical applications might be, but we still have an enormous amount to learn about the universe. I think we know only a very small fraction of what we ultimately will know. And some of the things we learn in the long and short term may in fact turn out to be practical. But it is hard to sell astronomy on the grounds of being practical. I think it is not unlike music and art. It is nice to live in a civilization that will support artists, and will support musicians, and it is equally nice to live in a civilization that will support astronomers.

J. W.: Well put. And if you had an opportunity to get on a space ship to go to a planet or possibly outside the solar system, would you take that opportunity? Or are you content to stay here?

V. R.: I think I am content to stay here. It will take a long time to get outside the solar system, almost a lifetime. It might be interesting to go to one of the relatively nearby planets. I think if someone seriously offered me a chance to go to Mars, I would probably say yes.

J. W.: Let's turn to your fifth and final choice today, and it is Bach's F Orchestral Suite No. 3. Why are you taking this?

V. R.: If I am really to be on a desert island, it seemed impossible to be there without some Bach. Out of the enormous difficulty of choosing one, I chose this just because I think it is very pretty, and it is something I could listen to over and over and enjoy it continually.

J. W.: Dr. Rubin, you have a rather nice suggestion for the general public. Why don't you make it now?

V. R.: It seems to me that our forefathers and foremothers were perhaps luckier than we are because they lived outside enough to be much more familiar with the sky than are most people today. And I think that if somehow we could get the public to go on a very dark night out into the country and to look at the Milky Way, most of the children would be overwhelmed. When I talk to elementary school and junior high children, I very often begin by asking them if they have ever seen the Milky Way. And these are children in Washington, D.C. And the answer is universally no. If we were to designate a month as the month of the Milky Way, and get people to drive 60 miles from their city out into the country and just spend a few hours in the early evening looking at the Milky Way, they could understand that they are looking at our galaxy. That the brightest part of the Milky Way that they are seeing in the south is toward the center of our galaxy. And I think we could increase not only their awareness, but their interest in science. People would see that these views are really spectacular.

J. W.: It is exciting and I agree they are missing a wonderful experience if they have never seen the Milky Way at its full glory. Dr. Rubin, on a desert island you are allowed a luxury item and a book, other than the Bible and the complete works of Shakespeare, which you are already given. What book would you take?

V. R.: I actually thought about this, and my first thought was that I would ask for a dictionary, because my husband and I have become addicted to doing the Sunday New York Times crossword puzzles, and I thought if I won't have a newspaper, at least I could make up my own puzzles. But, I decided that is not what I would ask for. Instead, I would like a book from which I could learn about the shells and the fish and the flowers and the trees on the desert island.

J. W.: What a good idea. And your luxury item?

V. R.: My luxury item, of course, would be a telescope. Even a small telescope, even if it had to be a pair of binoculars, but probably a small telescope.

J. W.: Very well, you shall have it. Dr. Rubin, thank you very much for joining me on the island today.

January 26, 1992

Dear Ms. Vera Robinson,

My name is Anne Duhamel. I am a fifth grader at Hanover Middle School, in Hanover, Massachusetts. I am doing an independent study project on the Evolution of the Universe for Sage, Shared Approach to Gifted Education.

I would appreciate if you could give me your opinions on this subject. My major research question is, will the Cold Dark Matter theory of the universe be disproven? If so, will the Big Bang theory be disproven also?

Could you please answer the following questions?

Do or did you believe in the Big Bang theory? If so, what is your reason?

Do you believe that the Cold Dark Matter theory is going to be disproven? If so, do you believe that it is going to be disproven by the new findings that there are clumps of matter in the universe and that matter was not distributed evenly after the Big Bang?

Do you believe that the new found evidence is going to show that galaxies were built long before the Big Bang theory suggests? Will this disprove the Big Bang theory?

If you believe that the Big Bang theory is going to be disproven, what theory do you think will take its place?

Do you think the Big Bang theory will be able to survive without the Cold Dark Matter theory? If there isn't any Cold Dark Matter, what is there that holds together the universe?

My project is due by the end of March. Would you please take the time to answer my questions? Thank you for all your help.

Sincerely,

Anne Duhamel

Carnegie Institution of Washington
Department of Terrestrial Magnetism
February 18, 1992

Dear Ms. Anne Duhamel:

Your letter concerning your fifth grade project on the Evolution of the Universe has reached me, and I am enormously impressed by your questions. You must have a very good science program, and a wonderful science teacher. Most of your questions are very difficult to answer, but I will do my best.

Do I believe in the Big Bang theory? "Believe in" is not the way astronomers would phrase it. Astronomers make observations which we then try to understand in terms of an overall model, a theory. If later observations cannot properly fit into the model, then the model must be modified, or even discarded in favor of a new one that better takes account of all the observations. So yes, I believe that the Big Bang model of the universe is better at present than any other model to account for the observations, especially the observation that space is expanding and carrying the galaxies with it.

At present, astronomers do not know enough to get very accurate ages for stars, and hence for the galaxies in which they reside. I am not worried that some stellar models make the stars older than the universe, for both ages are imprecisely known. Many astronomers currently are working on observing programs and on theoretical models which will some day result in more accurate ages.

The history of science teaches us that many theories must ultimately be modified, so it is likely that some of our current ideas concerning the Big Bang are wrong. Physicists learn about atoms, molecules, radiation, and matter in our earth-bound laboratories, and hence know about matter under conditions of temperature and pressure not too different from those in your school room. But at the origin of the universe, according to the Big Bang model, the temperature and density were unimaginably high. So we are not certain how well our "laws of physics," that is, our description of how matter behaves, are applicable to the conditions present in the early universe. If you become an astronomer when you grow up, this would be a good problem for you to attack.

If the Big Bang is disproved, what will take its place? That's too hard a question to answer, for it depends on which way the observations show that the theory is not a good fit to the universe we live in. But it will be a real advance for science, for it will mean that we will have learned more than we know at present, and that some of that new knowledge is a surprise.

Astronomers devised the Big Bang model of the universe long before we knew that most of the matter in the universe is dark. So yes, the Big Bang model can survive even if Cold Dark Matter is not an accurate description of that 90% of the matter in the universe which is dark. What is the dark matter if it is not Cold Dark Matter? We'll have to wait for observations, by astronomers in observing the universe, and by particle physicists in their laboratories, to answer this one. At present, "We don't know" is the only honest answer to what the dark matter is.

How did you learn to think about such important, difficult questions? You would make a good astronomer, for being curious about the universe is one of the most important qualities for being a scientist. That, plus a good imagination, so you can try to imagine what the universe is like, and then devise observing programs to try to get answers and see if your ideas are a good description of the universe.

You are a lucky young lady, for you will someday know what kind of material constitutes the dark matter. But if you are especially lucky, you may puzzle over some new questions we cannot now even imagine asking. Over fifty years ago, when I was your age, no one thought of asking questions about dark matter, for knowledge of dark matter in the universe was virtually nonexistent. So don't be afraid to dream up some really far out questions. Those questions may turn into the science of your generation.

Best wishes to you for an exciting future in science.

Sincerely yours,

Vera Rubin

Where in the World Is Berkeley, California?

1996

I n case you've forgotten, today is May 17, 1996. On May 17, 1948, years before you were born, I graduated from Vassar College. I remember details of that exciting day: the damp weather, my family and friends who joined the celebration, and the fun. But certainly I do NOT remember who delivered the commencement address. So this is a warning to me, and to you.

The invitation to address you tonight came while I was preparing to go observing at Kitt Peak National Observatory, to study orbits of gas and stars in galaxies. And on several disappointing rainy nights, I wondered what you might like to hear on this momentous day in your lives. I wondered if you realized how long is your past, and how much more there is in your future. I remembered a Peanuts cartoon that my family likes. Lucy is saying to Charlie Brown, "on the oceans of the world are many ships, and some of them carry passengers. One of the things the passengers like to do is to sit on the deck and watch the water. Some of the passengers like to face forward, so they can see where they are going, and some like to face backwards, to see where they have been." And then Lucy asks Charlie, "On the ship of life, which way are you going to place your chair; to see where you are going or to see where you have been?" And Charlie Brown replies, "I can't seem to get my chair unfolded."

Well, my chair is OK, and tonight I am going to look backward, to tell you how you are connected to the early universe, and how the early universe connects to Berkeley, 1996. John Muir understood these connections when in 1911 he wrote in *My First Summer in the Sierra*, "When we try to pick out anything by itself, we find it hitched to everything else in the universe." So this is a fable for our time. I hope you will someday tell it to your young friends. But you will be luckier, for you will know more than my generation does, and you can correct our tale.

About 15 billion years ago, when the universe was an enormously hot, enormously energetic, enormously dense point, it inexplicably, to us, started to expand. When one-hundredth of a second old, the universe had a temperature of 100 billion degrees. As it expanded, it cooled rapidly, meanwhile forming the space we know as our universe. No atoms, no molecules, only elementary particles existed; these being continuously created and then annihilated. But within three minutes of the origin, the temperature would cool so that protons and neutrons could combine to form nuclei of hydrogen, and helium. Ultimately, in some region, the fluctuations in the hydrogen density were so great that locally the expansion was halted, and reversed, and the density increased still further. This is the real miracle, for from this higher density region a star would be born, and in this star the elements heavier than hydrogen and helium would be produced. These are the elements necessary to make the earth, to make California, and to make you.

Deep in the interior of this hot star, where the temperature was 100 million degrees, hydrogen nuclei fused to form helium, and helium fused to form carbon, and nitrogen, and oxygen and other elements. Three helium nuclei fused to form the nucleus of one carbon atom of 6 protons and 6 neutrons. And so, 10 billion years ago, this carbon atom acquired its family of 6 electrons. Later, no longer welcome in this aging star, these atoms were shed into the spaces between the stars. Nature recycled these elements, along with the much more abundant hydrogen and helium, into new generations of stars enriched with heavier elements.

And then, less than 5 billion years ago, when the universe was about two-thirds of its present age, something very special happened in an outer arm of one of the billions of spiral galaxies which make up our observable universe. A small cluster of protostars formed deep inside a cold molecular cloud, their gravitational collapse initiated by the shock wave from a supernova explosion of a nearby dying star. John Muir would have liked this. One of these protostars is special, for it will grow to be *our sun*. But star formation is a messy process, and a rapidly rotating disk of debris surrounded the star; the gas and dust left over from the supernova explosion and left over from the star's formation. In this solar nebula our carbon atom abandoned its nomadic existence for a stable orbit.

At a distance of about 100 million miles from the star, it was cold in this solar nebula, and iron and magnesium silicates crystallized. These crystals, some wandering on erratic orbits, collided and adhered to form rocky planetesimals. Our carbon atom suffered a similar fate: it collided and fused to the surface of one grain. Quickly, planetesimals collided, merged, and formed moon size objects. It took longer, of the order of 10 to 100 million years, to form the earth from the merger of the largest preplanetary bodies. At the final accumulation stage, collisions with small, planet-sized embryos produced giant impacts.

Thus, four and a half billion years ago, our sun had a family of planets. But this is a long tale, for the universe is old. We could not exist in a young universe, for it took time to build the elements necessary for life. As we reach the present, the time scale shifts from *billions* to *millions* of years. When I talk to young school children, I often ask how long it will take to count to 1 million. The answer is about 10 days. And then I ask how long it will take to count to 1 billion. And they usually answer "20 days." But of course that is not so—it will take about 30 years to count to 1 billion. So note the change from billions to millions as the tale goes on. Life and human history are young features of our earth.

We live in a galaxy of 200 billion stars and there are uncounted billions of galaxies in the universe. Do other stars in our galaxy have planets? Without doubt, yes. In the last year or two, several have been detected. Do some of these planets support life? Yes, because the laws of physics and biology are similar throughout the universe. Will those life forms resemble ours? That's the hard question. In some cases, probably not, for it appears that life on our planet came about through a succession of random events. Just as a single life is a random walk and not a computer program, so the evolution of life on earth has been a continuing random walk. This question of life elsewhere is one that your generation will answer. But now, back to the evolving earth.

Four and a half billion years ago, our sun had a family of planets. Earth was a molten planet, heated externally by the sun and by impacts of objects almost as large as the moon, and heated internally by radioactive elements. The view in Disney's Fantasia is appropriate. The temperature was high enough to melt the iron which made up about one-third of the primitive planet. In a catastrophic process, the iron sank and formed a core, and the lighter molten rock rose, and formed a magma ocean. Outgasing gave rise to a fragile atmosphere.

Life, in the form of bacteria and algae, originated early. Our carbon atom spent the next billion years in this sluggish environment, ultimately arriving at the surface where it formed part of the primeval crust. This primitive crust melted and solidified repeatedly, weathered by sun, water, wind, and ice. Ultimately this shell broke into a dozen large, rigid plates which moved over the lower layers, driven by convection currents from the heat below. Earthquakes announce the incremental steps of these plates;

volcanoes mark the boundaries where continental plates override ocean plates. Continent formation has been a continuing process on earth.

While continents were forming on earth, galaxies, made early in the history of the universe, were evolving. In regions dense with galaxies, the expansion of the universe was retarded and in some places even halted by their mutual gravitation. Stable clusters of galaxies evolved. The galaxy containing our sun and our earth and our carbon atom was caught in an outlying region of the Virgo Supercluster, one of the dense regions of the clumpy universe. Here, massive galaxies gravitationally captured less massive ones, and disk galaxies merged to form spheroidal galaxies.

But now it is time to form California. Five hundred million years ago, when the universe was already 95% of its present age, the continent of Gondwana existed on earth. Primitive fish swam in the surrounding oceans, and land plants and trees would appear shortly. Gondwana divided, reassembled, and ultimately separated into the continents we recognize today. The earth was already old, and mammals and birds inhabited its outer layers.

For most of the history of the universe, there was no California. Early North America ended on the west with the plains, over which vast rivers crossed on their route to the western ocean. Then piece by piece, parts started assembling the West Coast in a disorganized jumble. An island arc here, an isolated island there, a piece of continent elsewhere, came crunching in and adhering—docking is the geologist's term. Tens of thousands of earthquakes, "big ones," accompanied as growing pains. About 165 million years ago, in the high noon of the dinosaurs, an island arc like the Aleutians docked, thus doubling the size of California. In so doing it exposed its Mother Lode of gold. Thus was the future history of California commenced. Three or four million years ago, recent for geologic times, the Sierra Nevada range started rising. Volcanoes were prominent.

The Coastal ranges have a different history. Rising sluggishly as islands from the deep, they resemble a marine clutter. When they pushed upon the continent, volcanoes arose to spew lava upon Napa Valley. A great depression filled with runoff from melting glaciers formed San Francisco Bay. As the Pacific Plate dives down under the continental North American Plate, it slides north. San Francisco arose in the trench marking one region of this geologic activity, the San Andreas fault. And the Berkeley Campanile stands like a marker on the parallel Hayward fault.

And what of our carbon atom? Carbon, alone among all the elements, binds itself into the long chains necessary for life on earth. Homo sapiens arose on earth 1 million years ago, when the earth was 99.99% of its present age. For the last thousands of years, the history of our carbon atom is a rapidly changing one. Breathed in and expelled by humans, dissolved in streams and oceans and re-emitted into the air, transported around the world by birds, we know it as CO_2, carbon dioxide. And then, some months ago, a common miracle occurred, one that occurs billions of times a day. Our CO_2 molecule, floating in air on a sunny California day, brushed the leaf of a buttercup growing on the campus of the University of California. Absorbed by the plant, transformed by the sun, mowed by a gardener, fed to a cow, our atom floated in your glass of milk.

You drank the milk, the carbon atom entered your bloodstream, traveled to your brain, displaced a carbon atom, and took part in the thought process permitting you to pass your final exam. So without that single carbon atom, made in some star billions of years ago, you might have failed to receive your diploma today. See how lucky you have been?

This then is the answer to the question, "Where in the universe is Berkeley, California?

My students and post docs caricature me by saying "Let me give you a reference." So let me give you some references. For the early universe, Steven Weinberg's *The First Three Minutes*. For the history of the earth's formation and evolution, George Wetherill's *Formation of the Earth*, and *Understanding Earth* by Frank Press and Raymond Siever. And of course, John McPhee's *Assembling California*. A single carbon atom is featured in a chapter in Primo Levi's remarkable book, *The Periodic Table*. This is the chapter I read aloud to my husband, my children, children-in-law, and grandchildren on a snowy night in Jackson Hole, Wyoming, some years ago.

So that is your past. And now, you must turn your chairs to face the future. You are concerned tonight with more than the fate of atoms. You need jobs, admission to graduate schools, research support; you want a healthy planet, space, choices. Individually, you will be called by many names: spouse, partner, teacher, professor, writer, representative, president, CEO, doctor, judge, regent. Some will be called scientist. For those of you who teach science, I hope that you will welcome, as students, those who do NOT intend to be scientists, as well as those who DO. We need senators who have studied physics and representatives who understand ecology.

And for those of you who choose to be scientists, I have one piece of advice. *Don't give up*. Science is hard and demanding, but *each* of you must believe that you can succeed. It may seem unlikely tonight, but there is not one among you who cannot make important, major contributions to the world of science. At my commencement on May 17, 48 years ago, the probability that I would be addressing you tonight surely was zero.

Instead of advice, I offer my hopes for you. I hope you will stay alert and heed the words of Yogi Berra: "You can see a lot by just looking." I hope your lives will be filled with health and peace, that you understand there is much work to be done in the world and that many of you will choose to join with those who work and lead. I hope you will disdain mediocrity and aim to excel in whatever you do. I hope you will love your work as I love doing astronomy. I hope that you will fight injustice and discrimination in all of its guises. I hope you will value diversity among your friends, among your colleagues, and, unlike some of your regents, among the student body population. I hope that when you are in charge, you will do better than my generation has. In 1993, U.S. Universities awarded PhD degrees in Physics and Astronomy to a *total* of nine black Americans. You do better.

The world you face is very different from the world of science I entered. I believe, indeed I hope, that virtually everything you studied in physics and astronomy was in no textbook that I saw as a student. My granddaughter, a freshman at UC Santa Cruz, is evidence that two generations have passed. I was a very impatient, very young, nonestablishment student. I had a BA degree and a husband before I was 20. A husband who is here tonight. Two young children came to my PhD commencement. Our four children are scientists. My achievements in science came about because I knew what I wanted to do, and I found professional colleagues among helpful, gentle astronomers. I was never discouraged by others who were sometimes discouraging. Instead, I insisted on working on problems outside the main stream of astronomy so that I could work at my own pace and not be pressured by bandwagons. I do not offer this as an example for you, but only to show that there can be diverse approaches to science. *There must be.* I hope some of you will be able to devise your own paths through the complex sociology of science. Science is competitive, aggressive, demanding. It is also imaginative, inspiring, uplifting. You can do it, too.

I hope that if it rains on your nights at the telescope, you will be treated, as I was, to a magic carpet display of wild flowers as you descend the mountain. By your outstanding education you have set yourself among the privileged. Each one of you can change the world, for you are made of star stuff, and you are connected to the universe.

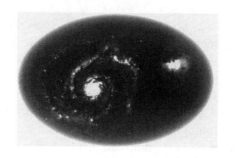

SOURCES

Most of the essays in this volume were originally printed or presented elsewhere. The publisher and the author wish to express their grateful acknowledgment to the following publications and publishers for permission to reprint previously published articles. The subject matter has not been altered to bring it up-to-date, but editorial changes have been introduced. Many technical photographs have been omitted; photographs included here are credited in the figure captions.

PART I: GALAXIES

THE PAST DECADE MEETS THE NEXT DECADE was originally published under the title "Stars, Galaxies, Cosmos: The Past Decade, the Next Decade" and is reprinted here with permission from *Science* **209**, 4 July 1980, pp. 64–71. © 1980 American Association for the Advancement of Science, Washington, D.C.

STRUCTURE AND EVOLUTION OF THE GALACTIC SYSTEM was published in *Physics Today* **13** (12), pp. 32–35, December 1960.

WALKING THROUGH THE SUPER(NOVA) MARKET has never before been published.

DYNAMICS OF THE ANDROMEDA NEBULA originally appeared in *Scientific American* **228** (6), June 1973, pp. 30–36. Reprinted with permission. © 1973 by Scientific American, Inc. All rights reserved.

THE PECULIAR GALAXY NGC 1275 was published under the title "Kinematic Studies of NGC 3351, 3115, and 1275" in the *C.I.W. Yearbook*, vol. 75, pp. 122–127 (1976), and is reprinted here with kind permission from the Carnegie Institution of Washington.

UGC 2885, THE LARGEST KNOWN SPIRAL GALAXY was published in *Mercury*, May/June 1980, pp. 78–79, and is reprinted here with permission from the Astronomical Society of the Pacific, San Francisco, CA.

NGC 3067 was published under the title "NGC 3067 and the Mass of the Universe" in the *C.I.W. Yearbook*, vol. 80, pp. 554–558 (1981), and is reprinted here with kind permission from the Carnegie Institution of Washington.

S0 GALAXIES WITH POLAR RINGS combines two reports published in the *C.I.W. Yearbook*, vol. 81, pp. 556–568 (1982) and vol. 82, pp. 581–583 (1983), and is reprinted here with kind permission from the Carnegie Institution of Washington.

A PAIR OF NONINTERACTING SPIRAL GALAXIES was published in the *C.I.W. Yearbook*, vol. 82, pp. 583–586 (1983), and is reprinted here with kind permission from the Carnegie Institution of Washington.

SOME SURPRISES IN M33 AND M31 was published in *The Outer Galaxy*, edited by L. Blitz and F. J. Lockman, pp. 247–252 (1988), and is reprinted here with kind permission from Springer-Verlag New York, Inc.

NGC 4550: A TWO-WAY GALAXY was published in *Mercury*, July/August 1993, pp. 109 and 126, and is reprinted here with permission from the Astronomical Society of the Pacific, San Francisco, CA.

PART II: TOOLS OF THE TRADE: TELESCOPES, A CATALOG, AND SOME MAPS

ASTRONOMY FROM HUBBLE has never before been published.

MOUNT WILSON OBSERVATORY: A BRIEF EARLY HISTORY has never before been published.

THE DEDICATION OF THE VATICAN TELESCOPE has never before been published. The quote from Diane Ackerman comes from *A Natural History of the Senses*, with permission from the author. © 1990 Vintage Books.

LETTER FROM CHILE has never before been published.

REMINISCENCES 1994: OBSERVING AT THE NATIONAL FACILITIES was published under the title "Reminiscences: Observing at the National Facilities" in *A Look at AURA 1994*, and is reprinted here with kind permission from AURA, Inc., Washington, D.C.

A REVISED SHAPELY–AMES CATALOG OF BRIGHT GALAXIES is a review originally published in *Sky and Telescope*, May 1982, pp. 478–479, reprinted here with permission from the publisher.

STAR CHARTS is a book review of *The Sky Explored: Celestial Cartography 1500–1899*, by Deborah J. Warner. It is reprinted here with permission from *Science* **210**, 7 November 1980, p. 632. © 1980 American Association for the Advancement of Science, Washington, D.C.

PART III: MATTER AND MOTION

THE LOCAL SUPERCLUSTER AND ANISOTROPY OF THE REDSHIFTS appeared in *The World of Galaxies*, edited by H. G. Corwin and L. Bottinelli, pp. 431–452 (1989), and is reprinted here with kind permission from Springer-Verlag New York, Inc.

THE ROTATION OF SPIRAL GALAXIES is reprinted here with permission from *Science* **220**, 24 June 1983, pp. 1339–1344. © 1983 American Association for the Advancement of Science, Washington, D.C.

DARK HALOS AROUND SPIRAL GALAXIES was published under the title "Evidence for Dark Halos Around Spiral Galaxies" in the First ESO-CERN Symposium on *Large-Scale Structure of the Universe, Cosmology and Fundamental Physics*, edited by G. Setti and L. van Hove, European Southern Observatory, Garching, Germany (1984), and is reprinted here with kind permission.

HOW MUCH DARK MATTER IS THERE? appeared in *Bubbles, Voids, and Bumps in Time*, edited by James Cornell, pp. 97–104 (1989), and is reprinted here with the permission of Cambridge University Press and © Smithsonian Astrophysical Observatory.

A CENTURY OF GALAXY SPECTROSCOPY was published in the *Astrophysical Journal* **451**, 419–428 (1995).

PART IV: THE ASTRONOMICAL LIFE: WOMEN IN SCIENCE AND OTHER HEROES, COLLEAGUES, AND FRIENDS

AN UNCONVENTIONAL CAREER was published in *Mercury*, January/February 1992, pp. 38–45, and is reprinted here with permission from the Astronomical Society of the Pacific, San Francisco, CA.

WOMEN'S WORK is reprinted from *Science 86* **7**, 58–65 (1986). © 1986 American Association for the Advancement of Science, Washington, D.C.

OPENING THE DOORS has never before been published.

SOFIA KOVALEVSKAIA: SCIENTIST, WRITER, REVOLUTIONARY is a book review of *A Convergence of Lives: Sofia Kovalevskaia: Scientist, Writer, Revolutionary*, by Ann Hibner Koblitz. It appeared in *The Mathematical*

Intelligencer **7** (4), pp. 69–73 (1985), and is reprinted here with kind permission from Springer-Verlag New York, Inc.

GEORGE GAMOW was published under the title "Recollections about George Gamow" in *Gamow Cosmology*, LXXXVI Corso, 1986, © Società Italiana di Fisica, Bologna, Italy, and is reprinted here with permission from the publisher.

E. MARGARET BURBIDGE is reprinted here with permission from *Science* **211**, 27 February 1981, pp. 915–916. © 1981 American Association for the Advancement of Science, Washington, D.C.

THE LEVIATHAN OF LORD ROSSE has never before been published.

GÉRARD DE VAUCOULEURS AND THE LOCAL SUPERCLUSTER was published under the title "The Local Supercluster" in *Gérard and Antoinette de Vaucouleurs: A Life for Astronomy*, edited by M. Capaccioli and H. G. Corwin, Jr., pp. 215–219 (1989), and is reprinted here with kind permission from World Scientific Publishing Co., PTE, LTD, Singapore.

RALPH A. ALPHER AND ROBERT HERMAN has never before been published.

DESERT ISLAND DISCS is published here with kind permission from WETA-FM, Arlington, VA.

DIFFICULT QUESTIONS has never before been published.

WHERE IN THE WORLD IS BERKELEY, CALIFORNIA? has never before been published.

Subject Index

ABOUT THE AUTHOR

Vera Cooper Rubin has always been curious. As a child, she asked endless questions about the stars in the sky. As a Master's degree student in astronomy, she wondered not only about the existence of large-scale motions of clusters of galaxies, but also about the general expansion of the universe. "Devising observations that will cause Nature to divulge a few more of her secrets has filled my professional life. Answers matter, but questions lead."

Raised in Philadelphia and Washington, D.C., Dr. Rubin attended local public schools. She obtained a B.A. degree in astronomy from Vassar College (1948) and an M.A. degree from Cornell University (1951). With the renowned George Gamow as her thesis advisor, Dr. Rubin obtained her doctorate degree from Georgetown University (1954). Her Ph.D. commencement was attended not only by her husband (Bob), but also by her one-year-old daughter (Judith) and three-year-old son (David).

The family and science were the two priorities in the Rubin household. Combining motherhood and astronomy in an amazingly complex ("it only looks easy") manner, Dr. Rubin then joined the faculty at Georgetown University and in 1965 moved to the Department of Terrestrial Magnetism of the Carnegie Institution of Washington. Two more children (Karl and Allan) were born within six years after she received the Ph.D. degree. All four children are Ph.D. scientists.

I have no easy answer for why our lives all worked so well. In large measure, of course, it is due to Bob, who did more than support me—he actively encouraged me to take advantage of professional opportunities that I thought were too complex to accept. Part of it came from a confidence, acquired at Vassar College, that I could choose the path that my life was to take. Part of it came from a determination to make it work.

For a young woman attempting to become an astronomer or physicist today, the road may be rougher than it was 40 years ago. Jobs are scarcer, competition is tougher, colleagues are more aggressive for fear of their own futures, and affirmative action has become a dirty word. Equal access is not yet available. Although more women are studying science, their numbers in research and academic positions creep up impressively slowly. For reasons that have deep historical roots, too few women have the opportunity to make the choices that I was able to make.

Many awards and honors have been bestowed upon Dr. Rubin. She has been a Phi Beta Kappa scholar, a Visitor to the Institute for Advanced Study, Princeton, N.J., a Distinguished Visiting Astronomer at the Cerro Tololo Inter-American Observatory in Chile, and a Chancellor's Distinguished Professor at the University of California, Berkeley. Dr. Rubin became the first woman permitted to observe at Palomar Observatory in 1965. She received the Dickson Prize for Science from the Carnegie Mellon University in 1994 and the 1996 Weizmann Women and Science Award. She was awarded the Gold Medal of the Royal Astronomical Society (London) in 1996. President Clinton awarded her with the National Medal of Science in 1993 and appointed her to the President's Committee on the National Medal of Science in 1995.

Dr. Rubin has served on numerous editorial boards, scientific councils, and review panels. She has also written almost 200 papers on such topics as the structure of our galaxy, motions within galaxies, and large-scale motions in the universe. Dr. Rubin also finds time to lecture both here and abroad, provide career guidance for young astronomers, and serve as a spokesperson and role model for women.